THE BLACK GIANT

A History of the East Texas Oil Field
&
Oil Industry Skulduggery & Trivia

JAMES M. DAY

Copyright © 2002

All rights reserved. Printed in the United States of America. No part of this book, either in part or in whole, may be reproduced, transmitted or utilized in any form or by any means, electronic, photographic or mechanical, including photocopying, recording, or by any information storage and retrieval system, without permission in writing from the Publisher, except for brief quotations embodied in literary articles and reviews.

For permissions, or serializations, condensations, adaptations, or for our catalog of other publications, write the Publisher at the address below.

Library of Congress Cataloging-in-Publication Data

Day, M. James
The Black Giant: A History of the East Texas Oil Field / James M. Day
p. cm.
Includes Bibliographical References
ISBN 1-893157-01-6
1. Oil fields--Texas, East-- History.   2. Oil fields--Texas, East--Miscellaneous.
I. Title
TN872.T4D34   1999
338.4' 78223982'0976414--dc21   99-21459
                                                                      CIP

Published by
BRIDGER HOUSE PUBLISHERS, INC
P.O. Box 2208, Carson City, NV 89702
1-800-729-4131

Cover design by The Right Type

Printed in the United States of America

10 9 8 7 6 5 4 3 2 1

TABLE OF CONTENTS

Preface		...i
PART ONE	**POOR BOY IN THE OIL PATCH**	
Chapter 1	Oilmen & Other Scoundrels	...3
Chapter 2	East Texas	...9
Chapter 3	Promoters & Lease Hounds	...15
Chapter 4	The Daisy Bradford No. 1	...21
Chapter 5	The Daisy Bradford No. 2	...25
Chapter 6	Scotland	...29
Chapter 7	The Daisy Bradford No. 3	...33
Chapter 8	Do Unto Others	...41
PART TWO	**THE BOOM**	
Chapter 9	The Boomers	...51
Chapter 10	Gusher	...61
Chapter 11	Wheeling & Dealing	...67
Chapter 12	The Black Giant	...73
PART THREE	**BIG OIL**	
Chapter 13	Big Oil v. Little Oil	...81
Chapter 14	Government — Texas Style	...91
Chapter 15	Hot Oil	...99
Chapter 16	Oil Patch Trash	...109
Chapter 17	Hard Times	...115
PART FOUR	**HISTORY**	
Chapter 18	Dollar Oil	...123
Chapter 19	Bottom of the Barrel	...135
Bibliography		...144

Sidebars — Oil Facts & Trivia

The First Texas Oilmen 3
Warning: Tall Tales Can Be Hazardous to the Truth....................8
Oil Lease Rentals & Royalties 10
Trendology..13
Why Oilmen Deal in Fractions...16
Poor Boying..19
How Not to Deal With a Used Oil Equipment Supplier.............. 20
The First Drillers 22
Oil Scouts or Spies?...24
Texaco Bashing...26
Shell: Too Dutch — Too Jewish?..30
The Woodbine..38
Would You Trust a Sinclair Oil Scout?.......................................40
An Unanswered Question...44
Creekology..46
Humble Oil Ain't So Humble... 53
Mineral Rights 56
Landmen 57
Texas Justice 59
Texas Billionaire Trivia 74
Assassin Trivia..75
Standard Oil Bashing...82
Big Oil's Mouthpiece 86
TRC & OPEC?..91
Producer's Oil Co. Legacy 94
Confused About Gallons & Barrels? & Oil Spill Trivia 96
Yellow Rose of Texas 100
Strippers..104

Posted Price 106
East Texas Hot Oil 106
East Texas Mud 110
Suspicious Characters 113
Texans in the White House 119
Cheaters 130
Who was J. Howard Marshall? 132
OPEC 136
Oil & Politics 138
Everyone's Favorite H.L. Hunt Tale 140

PREFACE

Until Alaska's North Slope bonanza in 1967, the East Texas oil field — *The Black Giant* — was the largest oil strike in the United States. Its discovery in 1930 changed the face of the oil industry — a con man found too much oil!

Several boring accounts have been written about the two legendary characters who emerged out of the East Texas oil field — **Columbus Marion "Dad" Joiner** and **Haroldson Lafayette "HL" Hunt, Jr.** With one exception, historical accuracy has been woefully absent, but even that work was discredited by a muckraker who claimed Hunt cooperated with the authors "on the condition mentions of him be favorable." A study of the works on the East Texas oil field reveals most are based on shaky recollections, romantic myths and glorified lies.

Reporters didn't follow Dad Joiner and H.L. Hunt around the muddy oil patch and hotel rooms to record their wheeling and dealing. Even if they did, the rascals would never have permitted witnesses to their shenanigans. As for press interviews, there were few and truth was not one of their strong points. However, they were not alone. "Big Oil" was also up to skulduggery behind closed doors. Therefore, there was no *film at eleven* to report what actually happened.

H.L. Hunt was a prolific but far from impartial autobiographer. Late in life the billionaire related hokey anecdotes recounting the life of a rugged individualist he desired people to believe that are as exciting as cold porridge. While many media hacks reported the eccentric's bunkum, that is not to say a few journalists have not offered interesting insights into the life and times of a brilliant, fascinating rogue and America's most successful independent oilman.

Nestled in the center of the East Texas oil field, Kilgore College is renown for the Kilgore Rangerettes, those pretty long-legged Texas girls who march in the Cotton Bowl parade on New Year's day. Its campus also houses the East Texas Oil Museum exhibiting the glory days when Kilgore was the heart of the oil boom. The museum's initial funding came from a Hunt Oil Co. subsidiary. Thus, it should come as no surprise that the museum's brochure recommends: "Pause for a moment in the Memorial Room to count the accomplishments of one of the many persevering oilmen, H.L Hunt." After you pass Hunt's ten-foot statue, the museum is enlightening and well worth visiting.

No one erected a statue to Dad Joiner. However, if you drive nine miles west of Henderson to Joinerville (it burned to the ground and is not on most maps), you will find a historical marker acknowledging Dad's discovery. For those not afraid of getting lost, take the winding unmarked road for three-quarters of a mile to the site of the first oil well in the East Texas field — the Daisy Bradford No. 3. Almost seven decades after its discovery it is still coughing up a few barrels a day for the Hunt Oil Co. When I last visited, I was disappointed to find the small marker (nothing else in Texas is small) erected by the State of Texas to commemorate Dad's discovery had been vandalized.

Dad Joiner never wrote an autobiography, although he quoted poetry and penned love letters to women he flimflammed. His cornball Bible and poetry quoting wouldn't go over today, but I'll bet he could think of something romantic and enticing to say if he was still alive and swindling old maids and widows. One cannot help but admire the old scoundrel they called *The Daddy of the East Texas Oil Field*.

This is not attempt to rewrite or correct history. It is a tale based on actual events with a whale of a lot of Texas literary license. While sitting in cafes and bars in Kilgore, Longview and Tyler, Texas, listening to old timer's yarns, there were times I couldn't keep a straight face. Damon Runyon could not have conceived the characters. Originally written as a movie script, *The Black Giant* was rejected by a major film studio because Dad Joiner was "an old crook without soul" and "most persons are ugly, ancient and/or evil." So much for Hollywood, except to ask: What has Hollywood turned out recently? (Few realistic films.) If you feel you must impersonate Saint Peter at the Pearly Gates, don't pass judgment unless you can transform yourself into a "poor boy" independent oilman — a wildcatter — or a hardscrabble farmer in Rusk County, Texas, in the 1920's and 1930's.

The story begins in an era called the Roaring Twenties, a myth Americans cherish. President Calvin Coolidge was claiming America was near its "triumph over poverty" and perceptively deduced, "When more and more people are out of work, unemployment results." Herbert Hoover campaigned on "rugged individualism." Today, we look fondly on the 1920's as the "good old days" because we weren't around to face the harsh reality. In truth, times were hard. Sixty percent of American families lived below what today we call the poverty level, a third didn't have electricity and thirty percent couldn't afford the luxury of indoor plumbing.

Preface

America's Depression of the 1930's arrived in East Texas in the 1920's. Hardest hit were the farmers and small towns waiting for years of drought to end. Meanwhile, "Big Oil" companies were dividing up the world's petroleum markets and fixing oil prices while grouse hunting at Achnacarry Castle in far off Scotland. Big Oil had never heard of the old bent cripple, Dad Joiner, but soon would. His "ocean of oil" would revolutionize the oil industry.

◆ ◆ ◆

As one who spent many years as a lawyer representing independent oilmen and companies and, on occasion, "Big Oil," it was inevitable I would catch oil fever and venture into the oil patch to promote and drill oil and gas wells. For fifteen years, I have taught oil and gas law at the Washington College of Law, The American University, which cannot assume credit or blame for my irreverence or opinions.

The Black Giant embraces oil patch jargon at the time of East Texas oil field discovery two early authors called *The Last Boom*. The hokey dialogue, based on anecdotes and myths, must be read as written — with tongue in cheek. Dubious tales are noted with a smile ☺. However, the actions of Big Oil, one legal historian called "sordid," are documented. The records of the United States and Texas governments weren't squeaky clean either.

No history of *The Black Giant* is complete without venturing into the workings of the oil industry and references to decisions of the Supreme Court of the United States and a country judge called the "Sage of East Texas." Shunning legalese, it is saturated with sidebars and footnotes chock full of trivia and descriptions of the petroleum industry in order to afford the reader a better understanding of the times and characters. (In another life, the author admits writing an oil and gas law text containing 786 footnotes.)

Dad's discovery engendered a ferocious battle between Big Oil and independent oilmen, bootleg or "hot oil," martial law, the first stirring of conservation of the precious black gold, legalized price fixing and the transformation of oil and gas law.

◆ ◆ ◆

The story begins in 1927 when Al Jolson made the first talking movie, *The Jazz Singer,* in which he bragged: "You ain't seen nothing yet."

Dad Joiner's *Black Giant* proved him correct.

PART ONE

POOR BOY IN THE OIL PATCH

If there are obstacles, the shortest line between two points may be a crooked line.

— Bertolt Brecht

1
OILMEN & OTHER SCOUNDRELS

In the summer of 1927, Dallas' population had barely reached 250,000. It was and remains second to Houston in population. It was named after George Miflin Dallas, Vice President under James J. Polk, rather than a genuine hero like Sam Houston, the first President of the Republic of Texas and general of the small Texas army that whipped Santa Anna and stole Texas from Mexico. Houston is also the oil capital of Texas, that's why its football team was called the Oilers. After the oil business went sour, the Oilers moved to Tennessee, where Sam Houston had been governor before he ran off to Texas.

In spite of playing second fiddle to Houston, Dallas had its share of oilmen — those who produce the black gold from deep in the earth. Dallas also had more than its fair share of oil *promoters* — those who promote oil deals and themselves. It's difficult to tell the difference because promoters foist themselves off as oilmen to gullible citizens and investors.

Dallas also had an animal called a *lease hound,* who made a living buying, selling and trading oil leases without giving a whit whether there was oil under the land. Texans didn't know what to do with natural gas, although it was heating and lighting homes in the Mid West. Columbus Joiner was a loose amalgam of oilman, promoter and lease hound and the only one in Texas who claimed there was oil in East Texas. This also made him a *wildcatter.*

FIRST TEXAS OILMEN

Most Texans hate to admit that the first oilmen in the Lone Star State were Yankees from Pennsylvania and Ohio and a smattering of Okies. (Few concede Sam Houston was born in Virginia.) Texans didn't know what to do with oil when they found it...they were more concerned with smelly cows and boll weevils in their cotton.

One of the first oil discoveries took place in Corsicana in 1884 while the town fathers were drilling for water. When they struck oil, folks were outraged. The black goo stuck to their boots and messed up their carpets and saloons. After the oil caught fire and almost burned down the town, a couple of Yankees from Pennsylvania arrived and slicked the locals out of the oil in a deal secretly financed by Rockefeller's despised Standard Oil Co.

Wildcatters are a special breed. They are gamblers... dreamers who search for oil in places no one else dares. This can make them very rich *if* they strike oil. More often, it leaves them penniless because these stubborn men look in the most Godforsaken places imaginable and seldom find the elusive black gold. East Texas was Godforsaken in 1927.

Dea England was as pretty as a Texas bluebonnet when the nineteen year old went to work for Columbus Joiner in the Gulf States Building, a rundown, low-rent edifice in Dallas. After wandering its dark hallways, she finally found a door with flaking black paint reading *C.M. Joiner, Esq.*

Unlike the stereotype slick Texas oilmen in ten-gallon hats portrayed by Hollywood (J.R. Ewing in *Dallas*), Columbus wore a wilted white shirt, tie and sported a straw boater. Nor was he a tall rangy Texan. Born on a farm in Alabama a few miles from the Tennessee border, he spent his early years in the Volunteer State and Oklahoma when it was a territory. His stature had been whittled down by rheumatic fever contracted as a child that left him bent over from the waist. The remnants of the malady never bothered the vigorous old man. Head down, he scurried about faster than most lanky Texans with his hands flailing behind as if he were swishing flies off his backside. Folks said that he found a lot of dimes on the street that way.☺

Columbus' smile set the young secretary at ease as it did with women of all ages. His boyish charm and a baby-smooth complexion, he attributed to a daily dose of carrots, seldom failed to pique their interest and his enthusiasm was infectious. Gossip around Dallas was that he was a dirty old man. However, the sixty-seven year old wasn't out to dazzle the teen age secretary with his sexual charms...at least not at the moment. In addition to $15 a week, he offered her a share in something far more intriguing...*an ocean of oil.*

Finding an ocean of oil would be expensive. Her first job was to prepare a list of potential investors, or in the promoter's vernacular, sucker lists. High on the list were widows. One of Dea's tasks was to periodically stop by the newspaper stand at the Adolphus Hotel to purchase yesterday's out of town newspapers for a penny or so each. What better source of new widows ripe for plucking was there than their husbands' obituaries? The obits provided an indication of the wealth of the deceased. She skipped notices of the passing of milkmen and shoe clerks. Wealth was left to the widows of bankers, industrialists, dentists and doctors, especially if the dentist or doctor was on his "suck...er... investors list." He followed the oil promoters' cliché: "The deal was so bad I couldn't sell it to a dentist and only half the doctors." Columbus seldom promoted deals to lawyers, the scoundrels always wanted to read everything.

According to Columbus, widows were not familiar with the world of finance and soon would become penniless, if left to their own devices. They could also fall into the hands of the unscrupulous. It was his duty to help them in their time of need. As he had practiced law in Tennessee and served in its legislature, he was fully cognizant of his fiduciary duties. Did not the Book of Jeremiah teach in Chapter Forty-nine, *Let thy widows trust in me*?

Spinsters, the *femme sole*, were not forgotten and were more easily aroused. Columbus is often quoted: "Every woman has a special place on her neck and, when I touch it, they start writing me a check. I may be the only man on earth who knows how to locate the spot." ☺ However he admitted that the checks weren't always good.

Less than two months of formal education were overcome by voracious reading. The Bible and poetry quoting con man held the record for the most books borrowed at the Dallas Public Library and was often seen bent over and scurrying towards his next mark with a Bible tucked under his arm.

H.L. Hunt went home to his wife most nights he was in El Dorado, Arkansas, during the 1920's. It was never explained how Lyda Bunker Hunt, a quiet Quaker and mother of four at the time, tolerated the "traveling man" and gambler. No one ever questioned his love for his wife and four children — Margaret, Caroline, Nelson Bunker and Haroldson Lafayette III, the later his favorite nicknamed "Hassie."

When in Shreveport, Louisiana, H.L. went home to his wife, Frania Tye Hunt, and two youngsters, one year old Howard and baby Haroldina. Born Frania Tiburski, she was but twenty-one when thirty-six-year-old H.L. met her on a business trip and they went through a quickie marriage. Either H.L. was cheap or short of cash when he purchased Frania's wedding ring in a pawn shop.

Shreveport, an oil town only 100 miles from Eldorado, was good reason for H.L. to call himself Major Franklin Hunt. Numerous accounts have been written about H.L.'s three wives and fifteen children, although H.L. mentioned little of Frania in his several autobiographies. Many lawyers became wealthy as a result of his alleged bigamy and fourteen living children squabbling over a couple of billion dollars.

Eventually, Fran's four children would get the short end of the inheritance. It was never explained why all their names began with *H*...Haroldina, Howard, Helen and Hugh, nor why Hugh was sometimes spelled Hue, other than H.L. only attended school for two months.

H.L. was thirty-eight and had been in the oil business six years when 1927 rolled around. He claimed to have run away from home in Illinois at the age of fifteen and spent the next decade wandering the West as a cowboy, lumberjack, mule skinner, sheepherder and gambler before settling down in El Dorado. He also worked as a dishwasher. As Haroldson Lafayette, Jr. was quite a mouthful to call a young boy, he was called "Junie." After he left home, the rough and tumble dishwasher shortened his name to H.L.

In El Dorado he gambled and speculated in cotton and real estate, also forms of gambling, before entering the oil business. A typical day in El Dorado was spent hanging around a pool hall, gambling casino or hotel wheeling and dealing, as he didn't have an office. Most mornings he could be found in the lobby of the Garrett Hotel dressed in a "spiffy" suit (one writer's description) and jauntily askewed fedora chomping on a giant panatella and diddling another slick promoter or potential investor. Nights were reserved for poker. In high-stakes games H.L. claimed he played under the name "Arizona Slim." In his early years he bragged that he got into the oil business after winning an oil well in a game of five-card stud. He denied the story later in life when his wife made him clean up his act.☺

One of the few good things about Prohibition was the Busch family built the elegant Adolphus Hotel in Dallas because it was against the law to brew their fine beer. The lobby of the Adolphus was one of Columbus' haunts, where he was often seen scampering about like a half-shut jackknife conniving with 325 pound Doc Lloyd. They were an unlikely pair. Doc often remarked that Columbus was the only interesting man he knew that did not smoke, drink alcohol or use profanity, all of which Doc did with exuberance.

Although seventy-three, Doc's infinite energy radiated. Like Columbus, he was a spell-binding con man. Dr. A.D. Lloyd's real name was Joseph Idelbert Durham. Some said he practiced medicine, others that he was a pharmacist from Cincinnati. He had also worked as a U.S. Bureau of Mines chemist and ran Dr. Alonzo Durham's Medicine Show, peddling medicines extracted from oil.[1] Most folks agreed Doc went under several names because of the six wives and numerous children he left in his travels.☺ There were also a few wags who claimed he wrote a lot of bad checks.

---

[1] Don't laugh, your mother probably rubbed Vaseline (petroleum jelly) on your butt when you were a baby. If you must laugh, laugh at the U.S. Bureau of Mines bureaucrats who had predicted eight years earlier that the United States would run out of oil in precisely nine years and three months.

The cohorts had muddled through several misadventures in the oil patch in the past. In Seminole County, Oklahoma, they drilled a wildcat well 3,150 feet. If they had the money to drill 150 feet deeper, they would have hit a gusher. In Caddo County, Oklahoma, the wildcatting duo missed hitting an oil field by less than one-half mile.

Doc looked the part of an adventurer in his habit of knee-high laced boots and jodhpurs topped off with a Mexican sombrero or cowboy hat. He told children and anyone else who listened that he always wore a hat to cover a scar buried under his white curly hair "ever since one of Pancho Villa's Mexican bandits split my head like a watermelon with a four foot sword." ☺ Doc went from play to play searching for gold and oil and always had an explanation for his great successes that didn't work out in the end: He left Mexico after finding a mother lode, but left for the Yukon when he woke up one morning and realized he was spending more time fighting bandits than digging for gold. He found nuggets as big as his fist in the Yukon, but it was so cold that he couldn't even get Eskimos to dig the gold out of the ground. ☹

WARNING

Tall tales can be hazardous to the truth. Yarns acceptable in the oil patch that will elicit smiles are designated ☺. Whoppers even your idiot brother-in-law wouldn't believe and will be met with frowns by oilmen are denoted ☹. The following examples of claims made by H.L. Hunt are furnished for guidance:

    1. As a young man I was a moderate drinker.☺ (As an old curmudgeon, he swore off booze.)

    2. When I was in my twenties, a group of men saw me stripped to the waist and urged me to challenge the black world heavyweight champion, Jack Johnson, believing I was the "white hope." ☹

    3. I am the best poker player in the world.☺

2
EAST TEXAS

East Texas had few paved roads. The dirt streets of the largest towns in Rusk County, Overton and Henderson, were mired with mud when it rained. However, it never rained when the farmers needed it. Long hot summers of drought weathered crops of cotton, corn and sweet potatoes. Stunted loblolly pines, scrub oaks and sweet gums provided little shade in the heat of the summer.

Rusk County wasn't bothered by sightseers. Tourists didn't flock to neighboring Gregg and Smith Counties either. The main roads between Dallas, Houston and Shreveport bypassed them. Paint peeled off most houses. The only signs of life around the shacks were half-naked children playing in the dust, razorback hogs rolling in the trash and mongrel dogs that barked as you passed, if it wasn't too hot. Barns and bastardized "yard cars" were the only landmarks.

In the fields white farmers and a sprinkling of black sharecroppers scratched the dry soil in the blazing sun with the help of mules. The farms were mortgaged and the outlook for paying off the bank looked bleak, which had its good points. Banks didn't foreclose because they already had taken more land under default than they could handle. "Coloreds" joked they were better off, the banks wouldn't lend them money, so they didn't get into as much debt as the white folks. Everyone knew what poor dirt and dirt poor meant. Yet, these hardy folks waived when you passed.

Doc's Buick spun a rooster tail of dust as it whisked along at a brisk forty milers per hour towards nowhere. Nowhere to Doc was arid land having no speak-easies, oil or road signs. Over a dozen dry holes, appropriately called "dusters," had been drilled in Rusk County in the past two decades. Doc and Columbus didn't need road signs, there was always the infinite patience of a stubble-bearded farmer who would lean on his hoe and point the way two or three times to make sure you "git thar."

Daisy Miller Bradford was a widow in her mid fifties. Childless and living alone on a 975 acre farm, she was lonely and craved attention, but no fool. Her doctor husband left her "quite comfortable," allowing her to maintain one of the few freshly painted farmhouses in the community. Unlike most women dealing with Columbus, when irritated by the old diddler, her tongue had a sharp bite. Daisy was witty, vivacious and as stubborn as an Arkansas mule.

The widow was crucial to Columbus's scheme. Her farm sat in the center of 5,000 acres he had leased during the past two years and she had surrendered one-quarter of the standard one-eighth royalty her neighbors received for his promise to drill the first well on her land. Daisy was rankled when Columbus and Doc arrived. Columbus was late making his delay rental payments of 50¢ an acre and he would have to call on his charm after the bumpy ride from Dallas to Overton.

OIL LEASE RENTALS & ROYALTIES

Oil & gas lease delay rentals are payable during the primary term of lease to allow the operator to delay drilling. The failure to pay a delay rental or drill automatically terminates the lease.

After an operator strikes oil or gas, the lease enters its secondary term for "so long thereafter as oil or gas is produced" and royalties are paid instead of delay rentals. For Columbus' commitment to drill the first well on her lease Daisy was to receive a 3/32 royalty instead of 1/8.

It was reported by an unreliable source that Columbus told Daisy: "I apologize for my untimeliness, but a voyage to find an ocean of oil is an arduous journey, such as the one taken by my namesake, Christopher Columbus. However, I now have my trusted navigator, Dr. Lloyd, and, if you have the faith of Queen Isabella, we shall set sail for our ocean of oil." ☺ In truth, Daisy knew there was no one else crazy enough to pay her annual delay rentals of $437.50 for land laying fallow, but that didn't meant she didn't crave an oil well and royalties.

Walter and Leota Tucker ran a general store in Overton. Both were leathery Texas stock — hard working, honest and, above all, dedicated. Texans can be a dedicated bunch. Remember the Alamo? In his slow lilting drawl, Walter looked every man in the eye, bearing the gospel of his words and demanding the same. Behind her boundless energy and captivating schoolgirl smile, Leota hid her shrewdness. Their children, Beverly, Mary and John inherited their Texas genes and dedication.

The Tuckers were firm believers in God and Columbus Joiner, but dared say that only Columbus would be the salvation of East Texas, as their eight years of praying for rain and prosperity had gone unanswered. What the Tuckers could not provide financially, they more than made up for in spirit and in kind. Many nights Columbus spun his tales of an ocean of oil while enjoying Leota's cooking. A storage room in the back of the store served as the wildcatter's East Texas office. Surrounded by canned goods at an old roll-top desk under a single light bulb, Columbus drafted oil leases and wrote romantic and business letters and romantic business letters.

J. Malcolm Crim knew there was oil under the family farm. He ran Crim's general store in Kilgore, a town of 700. A pillar of the business community, he never said anything unless it was important and rarely removed a dangling corncob pipe from his mouth. However, being a respectable businessman didn't mean folks failed to laugh when he told them how he knew there was oil under the parched land. This is but one version of the tale...

One sweltering summer day while Malcolm was visiting his in-laws in West Texas, he decided to escape his mother-in-law's glare and check out the town. Like most West Texas burgs, it only took ten minutes. To get out of the broiling sun, he slipped inside a Gypsy fortuneteller's tent and plopped down 50¢.

The shrewd Gypsy gazed at his palm and told him he lived in a big house on a farm, which anyone could have guessed — most everyone in rural Texas lived in a farmhouse and men in suits lived in big farmhouses. She also said there was a creek on the farm. That didn't take a crystal ball either — few farms could exist without water from a creek. However, she also saw a railroad crossing the farm. That sold him — not many farms had a railroad crossing it. But, that was just a warmup, Gypsies have to predict the future. Why else would they be called fortunetellers? She told him that there was oil under his farm...and the neighboring farms.

For many years, Malcolm Crim offered oil leases on thousands of acres around the family farm, but all the oil companies, small and big, laughed. There was no oil in Rusk county. Even Columbus Joiner's worthless oil leases were over twelve miles from his farm.

Geology is the study of the physical nature, history and structure of the earth's crust. It is an inexact science laden with correlations, unconformities and guesswork because the rocks and minerals are deep underground where no one can see them, unless there is an outcrop of a valuable ore or oil is seeping to the surface.[2] Even today's geologists and geophysicists, with seismographs, magnetometers, gravity meters and Geiger counters, cannot be sure if "thar's gold [or oil] in them thar hills." Oilmen will tell you that you never know if there is oil below until you drill. John Galey, a turn of the century oilman who made and lost several fortunes searching for oil and distrusted Ph.D. geologists was of the opinion: "The only reliable geologist is Dr. Drill."

It's common knowledge among oilmen, if you ask two geologists a question, you'll get four answers.[3] A layman wouldn't understand them anyway. They use big words like Cretaceous, Jurassic, Devonian and Precambrian. However, geologists know a lot of interesting facts of little or no importance except to themselves, such as the Pennsylvanian Period (Carboniferous) started 320,000,000 years ago

---

[2] Please, no nasty letters from geologists.
[3] See footnote 2.

(give or take a few years) when cockroaches first appeared. One spoilsport geologist ruined the movie *Jurassic Park* for his children by telling them that *Tyrannosaurus Rex* lived in the Cretaceous Period, not the Jurassic Period.

Rockhounds, as geologists are called, wander around looking at fossils and rocks and can tell you whether a rock is igneous, metamorphic or sedimentary. Sedimentary rocks are where you're more likely to find oil. One of the big words geologists love to tease you with is *anticline*, which is nothing more than an underground hill or dome. If the anticline contains sedimentary rock and is topped with an impermeable cap rock to keep the oil or gas from escaping, a geologist might tell you it's a "good prospect" to drill. He's not going to guarantee there is oil or gas in the anticline unless he's involved in skulduggery...like Doc Lloyd was when he wandered around East Texas.

As every reputable geologist (one who has not been caught in any hanky-panky) believed there was no oil in East Texas and over a dozen dry holes had been drilled in Rusk County, Doc — the former pharmacist and snake oil salesman — was going to have to write one hell of a intriguing geological report in order for Columbus to attract investors.

TRENDOLOGY

Doc Lloyd, experienced medicine show hustler, once professed to be trendologist — one who mapped the oil fields of Pennsylvania, Ohio, California, Texas and Oklahoma to determine where the oil mother load was located. He swore the declinations or trends of all American oil fields pointed to Rusk County as the apex of the apex, the heart and mother of all oil fields.⑧

Daisy and many others believed in people called doodlebugs, who found oil with divining rods, X-ray vision and electrical devices hooked to car batteries, but few believed in trendology.

3

PROMOTERS & LEASE HOUNDS

Promoters and lease hounds are considered oilmen...well sort of oilmen. *Webster's* defines oilmen as a person working in an oil field or an executive in the petroleum industry. This incomplete definition limits oilmen to blue-collar working stiffs and white-collar big shots.

A lease hound is a person who buys a lease from the owner of the mineral rights or another lease hound and sells or trades it to an oilman, naive greenhorn or another lease hound. He often obtains a lease from a landowner for the payment of a cash bonus, generally based on the number of acres in a lease, plus the promise to pay a royalty on the value of the oil or gas produced, say, the standard royalty of one-eighth (1/8). If he acquires a 100 acre lease for a bonus of $1.00 an acre and sells it for $2.00 an acre, he pockets $100. During the East Texas oil boom, lease bonuses jumped from 50¢ to over $1,000 an acre in a matter of weeks. That's when you're apt to run into more lease hounds than an East Texas coon dog has fleas.

Lease hounds might also attempt to retain an overriding royalty of, say, 1/32 or 1/16, which means the purchaser must also pay the lease hound a royalty based on the value of the oil or gas produced. Lease interests are divided into royalty interests (the mineral landowner) and operating interests (everyone else, including lease hounds receiving overriding royalties). Many early lease hounds preferred cash and did not fool around with overriding royalties. They preferred to sell a lease for as much as they could get then go to a saloon for a drink and buy and sell more leases.

WHY OILMEN DEAL IN FRACTIONS

Oilmen deal in fractions except in California where they understand percentages and never do anything the way everyone else does. If you remember your third grade math, fractions can be confusing, but not as complicated as multiplication and remembering where to put the decimal point. That's why many lease hounds went to a saloon rather than bargain for a 1/32 or 3/64 overriding royalty.☺

Promoters are deal makers and have to know something about the oil business. If they don't, they will hire a real oilman to drill the well. Let's assume a promoter estimates he can drill a well for $300,000. He might sell 3/4 (75% if he's in California) of the lease for $400,000. In the end he pockets $100,000 to cover the cost of the lease and for his time and trouble and retains 1/4 in case he finds oil. Naturally, oil and gas deals can get a lot more complicated, especially if you have a slick old codger like Columbus Joiner promoting the scheme.

Columbus' $25 Rusk County Oil Syndicate certificates were intricately engraved to look like Banana Republic legal tender or miniature blue chip stock certificates. Similar to today's real estate hustles, they fit into a No. 10 envelope. The certificate stated it represented an interest in the well on 80 acres of Daisy's lease of 25/75,000 based on Columbus' $75,000 investment valuation he picked out of the air. If you recall, your arithmetic teacher made you compute fractions to their lowest common denominator; thus, the certificate owner held a 1/3,000 interest in the well.

The syndicate also included 500 acres Columbus selected from his 5,000 acres of lease holdings, which entitled the owner to a 1/500 undivided interest in the 500 acres. Hence, *if* he sold all 500 shares of the syndicate at $25 face value, Columbus would gross $12,500.

Your math teacher might comment that Columbus was adding

apples and oranges. Actually, there are two computations: the interest in the oil well on 80 acres and the undivided interest in 500 acres (undivided means you do not own an interest in one particular acre, but 1/500 of the lease).

IF, and it's a big *IF*, Columbus sold all 500 shares of the syndicate and *IF* he struck oil, he would be entitled to 62,500/75,000 interest in the well's production...broken down to it's common denominator...5/6! A hell of a profit *IF* you can find folks that gullible.

Simple math is important to the tale: Columbus was adding apples and oranges plus a few onions. There were also a couple of other big *IF*s which will become apparent, not the least of which is: Where will Columbus get additional money, *IF* it costs more than $12,500 to drill a well on Daisy's lease? Another big *IF* was: What happens *IF* Columbus can't sell all 500 shares?

Before drilling, Columbus had to raise the money. To entice people on his sucker list, he enclosed a copy of Doc's impressive report on the ocean of oil to be found in East Texas entitled:

> *GEOLOGICAL, TOPOGRAPHICAL AND PETRO-LIFEROUS SURVEY, PORTION OF RUSK COUNTY, TEXAS, Made for C.M. Joiner by A.D. Lloyd, Geologist and Petroleum Engineer.*

The first person Doc's report impressed was pretty Dea England. She had to fold hundreds of copies and stuff them into envelopes. The report included a map outlining the Overton, Johnson Creek, Joiner and Lloyd Anticlines and was affixed with a *genuine* United States copyright...so it had to be true. Doc lied when he said there had been thousands of seismic registrations and gushers in the area as the basis of the report. He made the entire survey up. The only possible explanations for his naming the Lloyd and Joiner Anticlines are that he wanted to name something after himself and his old pal, Columbus Joiner, or he was committing fraud.

Along with the report, Columbus included a cover letter from Doc asking Columbus to *continue* sending him core samples and cuttings, which should have raised eyebrows — Columbus had not started to drill the well. Where did he get cuttings and core samples? It was never explained whether it was a typographical error or Doc was careless when stating in the report that Columbus would hit the pay zone at 3,055 feet and his cover letter said he would strike the oil bearing Woodbine sands at 3,550 feet.

Times were hard. The receipts from investors barely covered Columbus' expenses, so he journeyed to Rusk County to con money out of the locals. When he stepped off the train in Overton he was dismayed to see a sign in the window of Tucker's general store: GOING OUT OF BUSINESS.

Walter Tucker explained: We extended credit to the good people of Overton...They don't have jobs and can't make a dime farming...They can't pay us...We can't pay for more goods... And the suppliers won't ship us any more goods until we pay cash or they will go broke just like us. What Walter called a vicious cycle, economists call the Depression.

Columbus' answer to the Depression was to give Walter a quarter interest (1/4) in Daisy's well for future services to be rendered and make him an oilman. With Doc's glowing report in hand, Walter hustled local businessmen and sold one-eighth (1/8) for $900.[4]

With $900 and a pocket full of $25 certificates, Columbus and Walter set off for Houston County, 100 miles from Overton and the nearest area to buy oil field equipment. Walter had proven he could be a promoter. Now, Columbus had to teach him how to be an independent oil man and "poor boy" a well. They were not Standard Oil and didn't have Rockefeller's millions. They had to operate on a shoestring and drill their well with bailing wire, spit and a prayer — poor boying.

Poor boying ain't easy. Poor boys can't afford new equipment, like Exxon and Chevron. The have to deal with another character in the oil business — the used drilling equipment supplier — who make used car salesmen look like Billy Graham.

---

[4] Fractions are easier to keep track of than decimal interests, which is another reason oil men prefer fractions. *WARNING: Do not attempt to keep track of the fractions from this point.*

POOR BOYING

Poor boying is said to have originated in Oil Creek, Pennsylvania, in 1862. The story goes that J.W. Sherman couldn't start drilling because he didn't have enough money to buy an engine to power the drill rig. Unable to afford a new steam engine, he traded a 1/16 interest in the well for a horse. When the horse died two weeks later, he traded another 1/16 interest for a secondhand steam engine. In order to buy coal for the engine, he traded a 1/16 interest for $80 and a shotgun.

Sherman struck oil. During the next twenty years, the well earned $1.7 million, which meant that Sherman was no longer a poor boy. In case you don't have a calculator handy, a 1/16 interest earned $106,250...not a bad deal for the fellow who traded an old sick horse for a share of an oil well in 1862.

Columbus obtained enough secondhand equipment to start the drilling. As he couldn't afford a sufficient size boiler to power the rig, he settled for two used small mismatched boilers, one which had been used to power a cotton gin for over twenty years, the roustabouts named "Big Joe" and "Little Joe." The drill pipe was rusty and bent, described by one driller as "a heavy streak of rust." Naturally, Columbus didn't pay all cash for the junk. He traded the used equipment dealer a few crisp, freshly printed $25 Rusk County Oil Syndicate certificates.

HOW NOT TO DEAL WITH A USED OIL EQUIPMENT SUPPLIER

In the early 1980s, I took my eighteen year old son with me to visit a used oil equipment dealer in Oklahoma I will call "Uncle Billy," because that was what everyone else called him. As we were sipping a Coca Cola, a ritual Uncle Billy insisted upon during trading, a poor boy operator showed up I will call "Clyde," because my mother taught me never to make fun of idiots and drunks.

Clyde needed a compressor for a gas well Uncle Billy insisted was worth $18,000. Two hours of negotiation, including one break while they slipped out to Clyde's Chevy pickup for a shot of Jack Daniel's, resulted in Clyde paying $8,500 cash plus a 1/16 interest in the gas well and his Chevy pickup for the compressor and a sixteen-foot speedboat Uncle Billy had taken on a previous trade.

After they shook hands and Clyde left, my son asked Uncle Billy if he was aware there was an eight inch hole on the other side of the boat hidden under a canvas.

Uncle Billy spit a wad of tobacco and nodded. "An' I reckon Clyde will larn 'bout it, too, if the damn fool tries to put it in the lake."

Lawyers call that *caveat emptor.*

4
THE DAISY BRADFORD NO. 1

Tom Jones desperately needed a job, like many other Texans in 1927. He claimed to have been a driller in West Texas and worked as a miner in Colorado. Columbus hired him as the driller. At least he had more experience than everyone else Columbus had assembled as a drilling crew.

A driller is the chief of a drilling crew and crucial in drilling a well. As a working boss, he can be found on the rig floor wrestling with the drill pipe. He is charged with the responsibility for the condition and depth of the hole, if the drill bit needs sharpening and whether the crew is sober when they show up for work.

A drilling crew is made up of roughnecks, who work on the drilling rig floor, and roustabouts, oil patch slang for common laborers. Tom had to be satisfied with the former owner of a general store, farmers and fifteen year old John Tucker, whose father made him an unpaid roustabout. In the Texas oil fields, the entire crew would have been called *weevils*, a derogatory term for inexperienced men. If you know anything about cotton, you know that boll weevils are pesky little beetles and not highly regarded. Young John Tucker had the weevil chores of stoking the boilers, sharpening tool bits and catching catfish for the crew's supper.

Tom was paid $4 a day and promised $2 a day in oil lease interests *if* they struck oil. A major drawback in taking part of his pay in oil leases was the poor boy promoter determined the location and value of the oil leases. Drillers who worked poor boy wells were often gamblers or hard up for a job. Tom was both. When Columbus told him he had to drill to the Woodbine sands where they hoped to find oil at 3,550 feet, it's likely that Tom didn't have the money to get home. The drill rig was only designed to drill to a depth of 2,500 feet. What may have convinced Tom to stay was the exuberant weevils working for $3 day and determined to find Columbus' ocean of oil and the salvation of Rusk County. It also may have been Leota's cooking or that Beverly Tucker was pretty and would help with the cooking. The Tuckers planned to live in a tent on the site until Tom "brought in a gusher." Living next to the well would also eliminate the need for watchmen to keep an eye out for Big Oil company *spies*.

THE FIRST DRILLERS

America's first successful oil driller was William A. "Billy" Smith, a blacksmith and salt well driller who drilled Colonel Drake's oil well in Titusville, Pennsylvania, in 1859 to a depth of 69 1/2 feet.

The first known oilmen in America were the Seneca Indians of Western Pennsylvania. The Senecas collected oil from seeps and used it to caulk their canoes and as a medicine. Later, they sold oil to the white men as a cure for indigestion and toothaches and to grease their wheels.

In the fifth century BC, Herodotus reported that the Mesopotamians (Iraqis to those who forgot their ninth grade history) dug holes and scooped oil up in goatskin bags.

Where to drill the well? There are a dozen tales how the location of the well on Daisy's 975 acre lease was selected. Everyone agrees it was not the spot Doc Lloyd designated. Many believe Doc's choice was over one mile to the west. It is also agreed, if the well was located 300 yards to the east, it would have missed striking the East Texas field. An oft told anecdote is that Daisy insisted Doc's well location be moved several hundred yards because it was too close to her house and might dirty her washing hanging on the clothesline.☺

The Daisy Bradford No. 1

Progress was slow, but not sure. Much of Tom's time was taken up instructing the crew, repairing the rig and correcting the inexperienced crew's mistakes. A month behind schedule, the rickety pine and oak 112 foot derrick was finally completed. One of the weevils, Dennis May, whose real occupation was farmer and moonshiner, brought a jug to christen the well *The Daisy Bradford No. 1*.

The following day, they spudded the well (started drilling). With a green crew, secondhand equipment and rusty pipe most junk men wouldn't buy by the pound, drilling was slow. After drilling 100 feet, Big Joe and Little Joe, the 75 and 50 horsepower boilers, were barely able to power the drill through the hard rock. Columbus' inability to pay the crew's wages didn't help progress. Weeks went by with no drilling because Columbus failed to send money to meet the payroll and purchase supplies. Several *oil scouts* reported they had checked the well a dozen times and witnessed no drilling.

Some said it was inevitable. Six months after spudding, the secondhand pipe twisted and jammed. In oil patch jargon, they were stuck in the hole. They had only drilled 1,098 feet before surrendering to the odds.

Daisy, like many landowners who leased their land in those days, was shocked to learn that an operator had the right to use the water and wood on the lease for the drilling operations. Poor boy operators fired their boilers with wood instead of oil to save money. Although disheartened by his denuding her farm of trees, she handed Columbus a check for $100 and said, "Go raise the rest and get me an oil well."

OIL SCOUTS OR SPIES?

They weren't called *spies* in the early days. This little known oil industry position was euphemistically called *oil scout*. Their job was to obtain inside information on the drilling progress of other oil companies, such as the depth of the well, the formations penetrated and whether oil was found so their companies could reduce drilling costs and risks. They frequented card games, saloons and brothels to pick up information as the data was valuable and it was unlikely that the companies would release their drilling results. Oil scouts were not above sneaking around the well at night to steal drilling cuttings and peek at records. Actually, spies was a good name for the rascals.

5
THE DAISY BRADFORD NO. 2

As usual Columbus was physically bent and financially broke. He formed a new oil syndicate with another five hundred acres of leases, which boiled down to merely printing new $25 Rusk County Oil Syndicate certificates and mailing them to his sucker list. If everything went well, $25,000 would be sufficient to complete a well. Back to grammar school math: How many $25 certificates are required to raise $25,000? Answer: Lots. Forget your grade school math. Drilling a second well after reaching only 1,098 feet on the first well promised to hit pay at 3,550 feet did not instill investor confidence. When he couldn't sell the certificates at face value, he discounted them for up to 50% and sold oil leases until he had enough to start drilling a new well.

The determined and dedicated Tuckers were waiting for his return, including young John, who was paid $76 in advance to return to work, the first money the lad was paid by Columbus. One of John's first purchases with the huge sum was a new fishing rod to catch fish for the crew's dinner.[5]

---

[5] In their detailed account, *The Last Boom*, James A. Clark, a journalist, and Michel T. Halbouty, a geologist and petroleum engineer, report that John was paid exactly $76, but do not venture a reason for the odd amount. However, the author takes their word as gospel. Halbouty was President Reagan's advisor on energy matters during the transition from the Carter administration and recommended that the Department of Energy be abolished. Reagan should have listened to him.

One of the tales folks enjoyed telling about the Daisy Bradford No. 2 allegedly happened when Bill Osborne, the new driller, asked Columbus where he wanted the new well spudded. It was not an easy task to move a rickety 112-foot wooden derrick. By now the crew were no longer a bunch of weevils and their reply in unison was: "Downhill, any direction as long as it's downhill."☺

Money became harder to find than Dennis' five-gallon-a-day still in the pathless woods along Johnson Creek. Leota conned her brother, Wilfred, who had taken over the Tucker's general store, into extending credit for the crew's meals. Tools and repairs to the antiquated rig were paid for in $25 syndicate certificates sometimes discounted to less than $10. There was no telling how many were issued or who owned them as they circulated in Rusk County like money.

News that the Texas Company (now Texaco) hit a duster southeast of the Daisy Bradford No. 2 after drilling over 3,500 feet made money almost impossible to raise. When a Texas Company oil scout laughed at the crew and uttered the worse damning slur anyone can say to an oilman, *I'll drink every barrel of oil you can get out of that hole*, Dennis May picked him up by the collar and was about to clobber him when Walter Tucker stepped between them.

TEXACO BASHING

Dennis wasn't the only Texan who thought everyone from the Texas Company was an SOB. Texaco was founded in Texas as the Texas Company by Joseph Cullinan, a transplanted Pennsylvanian who was so scruffy and obstinate he was called "Buckskin Joe." After making the company worth a damn, Wall Street bankers stole the company from him and moved the Texas Company to New York.

Pennzoil, a Pennsylvania company that was once part of Rockefeller's Standard Oil Trust, but had the good sense to move to Texas, became miffed in 1984 when Texaco tried to muscle in on a deal Pennzoil had made to buy Getty Oil. Pennzoil sued Texaco in Texas and obtained the largest jury award in history — $11,120,976,110.83. Apparently, the jury of Texans still thought folks from Texaco were SOBs. In 1987 Texaco also gained the dubious distinction of becoming the largest corporation to file bankruptcy in United States history. Pennzoil had to be satisfied settling for $3 billion.

In order to survive, Texaco made a deal with Saudi Arabia to sell one-half interest in its three refineries and gasoline stations in the 33 eastern states.

Some said it was inevitable. (Sound familiar?) A year after spudding the Daisy Bradford No. 2, the corroded drill pipe twisted off in hole after drilling 2,518 feet. The drilling appeared even more pathetic when compared to Texaco's nearby well drilled over 3,500 feet in six weeks. The fault couldn't be laid to the green drilling crew. The equipment was dilapidated and most of the time the well lay idle because Columbus failed to pay their wages. Two weeks of attempting fish the broken pipe from the hole were futile and Columbus couldn't afford the tools to make further attempts.

Bill Osborne, the new driller quit — he hadn't been paid in a month. The crew returned to their farms to eke out a meager living and feed their families. Walter Tucker packed up his family and returned to Overton to take a job as a bank clerk. Columbus went home to Dallas, broke as usual.

It rained that day, then stopped as quickly as the downpour had started and the sun broke through. The brief rain and the Daisy Bradford No. 2. were false hopes. Rusk County was still facing drought and poverty.

6
SCOTLAND

Scotland in the fall is breathtaking. Grouse hunting and fishing are at their best. Achnacarry Castle was an ideal place for fun-loving oilmen to enjoy the outdoors, camaraderie and chat about oil prices in 1928, in case you are wondering what Scotland has to do about oil.

The fellows gathering in Scotland were not your run-of-the-mill good ol' boys from the Texas oil patch. The host was Sir Henri Deterding, the head of the Royal Dutch Shell Group (Shell). His guests included Walter C. Teagle, Chairman of Standard Oil of New Jersey (Exxon); Colonel Robert W. Stewart, Chairman of Standard Oil of Indiana (Amoco); William Mellon, President of Gulf Oil (later swallowed up by Chevron); and Sir John Cadman of Anglo-Persian Oil Company (British Petroleum.)[6]

There are a few things you should know about these chaps representing "Big Oil." The purpose of the meeting was to agree to stop competing in price wars, fix world petroleum prices and engage in other skulduggery. Shell was teed off at Exxon for buying cheap Russian oil Shell claimed it had an exclusive franchise to buy that Exxon was selling at bargain basement prices in Europe, and Shell and Mobil were engaged in a price war in the Far East. The foreign companies thought United States antitrust laws were a nuisance and the Americans were off their rocker for not fixing prices. Fortunately for the American companies, the Webb-Pomerane Act of 1918 permitted them to take part in antitrust conspiracies outside the United States.

---

[6] Current oil company names are provided to facilitate brand loyalty or brand treachery identification at the pump.

British Petroleum wasn't concerned about conspiracies and combinations in restraint of competition, it was 51% owned by the British government. Winston Churchill, then Chancellor of the Exchequer (the American version of the Secretary of the Treasury), was in favor of higher profits going into the Exchequer. In 1914, as First Lord of the Admiralty, Churchill had purchased controlling interest in British Petroleum's predecessor (Anglo-Persian) for the British government so the British navy would have a secure oil supply. The government decided not to purchase controlling interest in Shell, which was 60% Dutch and 40% British owned. The Dutch royal family owned a big chunk of the stock and Deterding was an abrasive Dutch nationalist, so Shell was "too Dutch." Also, as Shell was financed by the Rothchilds and British interests were managed by Marcus Samuel, it was "too Jewish."

SHELL: TOO DUTCH — TOO JEWISH?

Marcus Samuel was a nice old Jewish man popular enough to be elected Lord Mayor of London, but wasn't considered a real oilman, so he wasn't invited to the grouse hunt in Scotland. His father had sold boxes and trinkets made from seashells. Either in honor of his father or a hang-up on shells, he called the company Shell Transport & Trading and named all Shell's tankers after seashells. He even named one the *Clam*. Shell's logo is still a scallop.

Henri Deterding combined Royal Dutch's vast oil resources in Indonesia, then known as the Dutch East Indies, with Marcus Samuel's trading company, which had been dependent on unreliable Russian oil. Deterding wasn't nice. He became too pro-German for Shell's board of directors in the 1930's and was forced out as managing director. He promptly divorced his Russian wife, married his German secretary and moved to Germany where he became a buddy of Adolph Hitler.

Amoco's Colonel Stewart was small potatoes compared to the others and contributed little to the plot. Amoco was more interested in Venezuelan oil and Stewart was concerned about a 1922 scam back home that was about to be unveiled. Stewart and a poker playing buddy, Harry Sinclair of Sinclair Oil, had set up a dummy corporation to purchase 33,333,333.33 and 1/3 barrels of crude oil at $1.50 a barrel. The odd volume of oil sold was the whim of A.E. Humphreys, the seller and discoverer the Mexia oil field in Texas, who thought the Texas-size deal should be rounded off to exactly $50,000,000. Sinclair and Stewart sold the oil to their companies, Sinclair and Amoco, for $1.75 a barrel. John D. Rockefeller, Jr., whose family still owned 15% of the stock in Amoco after the Supreme Court ordered the break up of the Standard Oil Trust in 1911, thought it was uncouth for Stewart to swindle his own company and was successful in removing Stewart as chairman. An $8,333,333.33 scam wasn't peanuts in 1922.

William Mellon of Gulf was well-connected. His uncle, Andrew Mellon, was Secretary of the Treasury under Harding, Coolidge and Hoover and a major Gulf shareholder. Uncle's art collection was so large, he built the National Gallery of Art to hold the part that wouldn't fit in his mansion. Uncle Andy also came in handy later as Ambassador to Great Britain when Gulf and British Petroleum settled their squabbles over who owned British-controlled Kuwait's oil riches, they finally split 50/50.

Walter Teagle, called the "Boss" at Exxon, knew the conspirators had too much oil and prices would drop if they continued to compete. He had recently completed a deal for Exxon and Standard Oil of New York (Mobil) to muscle in on the Shell's and British Petroleum's oil interests in the Mid East (with help from the United States government).

Although Texaco and Standard Oil of California (Chevron) didn't go fishing with the boys, they later joined the group that was to become infamous as the "Seven Sisters." The Sisters had to admit Compagnie Francois des Petroles (CFP) of France into the family because the French government owned 25%. Not as big as the others, CFP was a "stepsister." Together, the eight oil giants controlled over 90% of the world's oil exports and *all the oil in the Mid East.*

In the afternoons when they were not hunting and fishing, Big Oil drew up an agreement entitled "Pool Association," which became known as the "Achnacarry" or "As Is" Agreement when it came to light in 1948. The Truman administration did not release the report to the public until 1952 when Harry decided not to run for reelection, and even then withheld some of the dirt.

In effect, Big Oil agreed to maintain their world market shares "as is" and not compete against each other, to share facilities and exchange oil so it could be delivered by the shortest and cheapest route and fix oil prices. They also agreed to let the other companies in on the deal, including Russia. The communists proved they understood the capitalist system...they cheated like the other members when they could get away with it.

In order to avoid United States antitrust laws, Big Oil did not fix prices in the United States. Instead, the world price of crude oil was to be based on the "relatively stable" price of crude oil in the *Texas Gulf Coast* plus transportation costs from Texas to the point of delivery, which added more profits, as the tanker costs to Europe from Texas were higher than from the Middle East to Europe.

As the Achnacarry Agreement was secret, the only comment about the week Big Oil spent in Scotland came from Walter "Boss" Teagle of Exxon: "The hunting was lousy."

Needless to say, if Columbus Joiner found the ocean of oil he promised his investors and folks in East Texas, it would upset Big Oil's plans.

7

THE DAISY BRADFORD NO. 3

Walter Tucker's job in the Overton State Bank didn't halt his dream of finding oil in Rusk County. When there were no patrons in the small bank (quite often in a town with little money), Walter related to R.A. Motley, the bank's president, if Columbus had a "wall hook" to retrieve the broken pipe, he could start drilling again. As expected, Motley asked: "What the hell is a wall hook?"

Walter explained a wall hook was a fishing tool that is lowered in a drill hole and into a broken pipe. Expandable, it latches onto the inside of the pipe so it can be pulled out. And best of all, they could rent one to fish the pipe out for $100.

Motley, like Walter, believed oil would be Rusk County's salvation. He loaned Columbus the money for the wall hook. Walter also believed that Motley had a stack of Columbus' oil syndicate certificates in the bank vault and wasn't inclined to let the bank's shareholders know how gullible he was.

The nearest wall hook was owned by Ed Laster, a driller living in Shreveport, Louisiana. At forty, Laster was a proud man with twenty years experience in the oil fields. Veteran drillers hated to admit they owned a wall hook. It could be taken as a sign they were incompetent and had "lost the hole." He refused to accept money for the wall hook. It was tainted — 'taint his and 'taint needed by a good driller.

Columbus recognized Ed as capable of poor boying his well down to 3,550 feet, but Ed wasn't keen about working for Columbus. His experience of late wages and struggling with dilapidated equipment and rusty pipe had taught him hard lessons he didn't want to repeat. Also, Ed didn't believe the skinny meandering hole flooded with water could be saved. He told Columbus that he would take the job if another well was spudded. When Columbus argued that he couldn't afford to start over, Ed returned home. A week later Columbus called the driller. He had raised some money. Ed could spud an new well on a promise of wages of $6 a day in cash and $4 in oil leases, which would be worth a fortune *if they struck oil.*

Before spudding the Daisy Bradford No. 3, the rickety 112-foot derrick had to be relocated 500 feet to the nearest level spot over rough ground studded with stumps and boulders. If Ed had his druthers, he would have built a new derrick, but relented to Columbus' pleas to save money.

With a slow and steady pull from a tractor, Ed and two faithful roughnecks, Dennis May and Glenn Pool, carefully skidded the creaking, teetering tower over logs the first 300 feet. However, it was impossible to avoid all the boulders and tree stumps in their path. When the wooden derrick hit a rock, its flooring snapped like a pencil. After swaying for an interminable time, the tottering landmark became East Texas' version of the Leaning Tower of Pisa.

Unable to move the derrick without a new sill, Ed ordered Dennis to buy lumber for a new base. Dennis advised that was impossible unless Ed had $10. The lumberyard had cut off Columbus' credit. After a string of obscenities, Ed joined the laughter of the two roughnecks and told them to level the derrick where it stood. So much for Doc Lloyd's precise location. But, if there's an ocean of oil down there, what's a couple of hundred feet one way or the other?

1200 FEET IN TWO DAYS the *Henderson Times* reported. Dennis and Glenn were amazed what new pipe (well, almost new) and Ed's skill could accomplish. Daisy claimed she wasn't the slightest bit surprised.. She knew a good driller when she saw one. To assure Ed didn't leave, Ed lived in Daisy's farmhouse. Daisy moved to her brother's home in Overton so folks wouldn't think there was any hanky-panky going on, but visited *her* well almost every day.

That was the good news. Columbus' sucker lists dried up and mere dribbles of cash flowed for the drilling. The crew had cut down most of the trees on Daisy's farm for boiler fuel when drilling the first two wells. Without credit, only freshly cut green wood was available,

The Daisy Bradford No. 3

which made more smoke that heat. The crew's paychecks bounced. On occasion, Motley covered the overdraft on the crew's wages to keep the work proceeding. Local suppliers and machine shops lost faith in "Joiner's Folly" and refused to take the Rusk County Oil Syndicate certificates at any price and Ed had to repair the rundown equipment himself. Romantics claimed Ed was dedicated and determined not to leave until the job was finished.☺ Dreamers maintained he was certain there was an ocean of oil under the land.☺ Practical men said there wasn't another driller's job to be found east of the Trinity River and damned few jobs open west of it.

Green wood couldn't provide sufficient heat to fire the boilers and power the drill constantly, and the deeper rock was harder to penetrate. The crew would stoke the two mismatched boilers until they generated sufficient pressure to drill for a few minutes. When the pressure dropped, they repeated the process. Progress fell to a few feet a day, and weeks would go by when drilling halted because there was no money to pay the crew or buy green wood.

Daisy could tell if the crew was drilling from a mile down the road. Black smoke could be seen billowing from the boilers and drifting down the valley. Unable to purchase wood, Ed turned to burning used automobile tires salvaged from junkyards. With a short-handed crew, two or three men would run back and forth attempting to keep up the boiler pressure as they drilled. Then the inevitable happened: Ed and one of the crew were stoking the boilers when Big Joe exploded, scalding them with steam. Badly burned, Daisy and her sister-in-law nursed Ed back to health in the Just Right Hotel owned by her brother.

Through a bandaged face, Ed told a teary-eyed Daisy: "We ain't got no good pipe...We ain't got no boilers...We ain't got no money...And we owe every supplier between hell and Houston...I'm going home."☺

There was only one thing for Columbus to do: organize another syndicate! This time he would go big time — $100 certificates! Each certificate bore a 1/300 interest in 80 acres surrounding the well and entitled the owner to four acres out of a 320 acre leases reserved for future drilling. **Stop!** Back to grammar school math. Dea England calculated that 300 certificates times four acres would require 1,200 acres, but she had faith that Columbus knew what he was doing, even when he sold several leases more than once.

Cherished bunk...a myth:[7] One story goes that Columbus struck a bonanza on Rusk County's back roads. As Columbus couldn't drive, Beverly Tucker drove him around East Texas. Beverly would not say how much he collected and said propriety would not permit her to repeat names. When she arrived to pick up Columbus one evening in front of a dimly lit porch, he was kissing a spinster's hand and she heard him say: *O queen of queens, how far dost thou excel, no thoughts can think, nor tongue of mortal tell.*

Not only could Columbus quote Shakespeare, but early one morning when Beverly picked him up at the home of a widow a mile outside Overton, Columbus looked plumb tuckered out and quoted an obscure poet named Bourdillon when he climbed into the car. *"The night has a thousand eyes, and the heart but one; Yet the light of a whole life dies, when love is done.* Whew, you know, sparking and financing an ocean of oil would be much easier, if I wasn't seventy. To the Overton State Bank!"

There was enough money in the bank to convince Ed Laster and the faithful roughnecks, Dennis and Glenn, to return to work, but it didn't last long. The Depression had hit nationwide. Banks were closing across the country. The stocks and bonds of Columbus' doctors and widows were worthless and their dividends dried up faster than leaves in winter. Soon they were back to burning junkyard tires. The crew became expert at handling the rusty pipe in the morning, which often jammed in the hole after sitting in water and debris overnight and had to be "unfroze." At times, Ed was glad to have the wall hook to retrieve the corroded pipe. But they were below 2,600 feet...deeper than they had ever been. However, they were still 950 feet from their target, the Woodbine sand, *if* Doc Lloyd was correct that the oil bearing Woodbine was at 3,550 feet.

---

7 Who could question a tale told the author by a ninety year old former roughneck from Kilgore, Texas, in 1995? His neighbor heard it from a fellow whose cousin overheard Beverly Tucker's best friend swear to her mother it was true. ®

The Daisy Bradford No. 3

With no funds Ed Laster was about to give up when Daisy told him Columbus had telephoned to ask Ed to take a core sample in a few days when Columbus was bringing prospective Dallas investors to the well. Ed knew Columbus had called Daisy rather than ask him directly. Core samples were only taken to determine the formation being drilled or if it was believed to contain oil, and he was 950 feet from the Woodbine. Ed was also aware that promoters had been known to salt core samples with oil.

Con men tell their story so often and so well that they believe it themselves. For five years Columbus had wandered across Texas, stopping in country stores, cafes and hotels, to tell anyone who would listen of striking an ocean of oil in the Woodbine. He would trace the Woodbine formation on a map of Texas east of Dallas, pointing where it hit the Mexia oil field...the Powell oil field...the Boggy Creek oil field...and into Louisiana oil strikes. The Woodbine had to be in East Texas. Store owners, hotel clerks, waitresses and farmers listened in awe. But most times he could only sell his $25 certificates discounted to $10 or $15, barely enough to pay for his hotel room and meals.

This time it was different. Investors from Dallas were not naive widows or simple farmers. The crew prepared for the big day, including the volunteer weekend roustabouts, Walter Tucker and R.A. Motley, who never doubted their visions of the ocean of oil. They had a stake in the well and the future of their county, so they lent Ed the money for the core barrel.

The prospects were not disappointed. Ed, assisted by Dennis and Glenn, pulled the core sample, as if they had worked as a crew for years, and laid it on the floor of the rig. The seventy year old promoter/actor smelled the rock cylinder and rubbed his hands along its course grains. Hesitating, as if deep in thought, he announced that it appeared they were almost on top of the Austin Chalk, so the Woodbine must be close by.

There was little for Ed to do but agree. He was part of the sham.

After Columbus and the prospects left, no crew member would bet against the con man taking their money.

THE WOODBINE

Geologists label each rock layer or formation underlaying the surface much like ham, salami, and bologna in a hero sandwich. In Texas, they confer cute names, such as Blossom, Cotton Valley and Smackover. Woodbine is a type of honeysuckle vine. (In Alaska, they named one after a whale: Beluga.) They also name them after heros, such as the Austin Chalk that overlays the Woodbine sand. Chalk is consistent with the Cretaceous period (65 to 144 million years ago, give or take a few years). Cretaceous means having the nature of chalk. However, all sediments deposited in the sea once covering East Texas and West Louisiana were not chalky. Many are limestone (organic materials from coral, mollusks, etc.), shale, (mud) and sand, such as the Woodbine. Under the right conditions, ancient ocean sediments might contain the remains of plant and animal life that have evolved into oil or gas over the millions of years due to heat and pressure. If you were buried 3,550 feet deep for 100 million years under heat and pressure, you would turn into a blob of oil or a belch of gas, too.

Money trickled in as a result of the impression made on the gullible prospects who had witnessed the core sample. Encouraged by the scam, Columbus convinced Ed to send him more telegrams advising that he was taking core samples. To the unsuspecting investors in the lobby of the Adolphus Hotel, core samples meant that the Daisy Bradford No. 3 was on top of the oil bearing Woodbine.

News of the only well being drilled in East Texas piqued the interest of oil scouts, whose job was to spy on the drilling progress of other companies. One of the first to visit the Daisy Bradford No. 3 was Don Reese of Sinclair Oil & Refining Co., America's largest and most noto-

The Daisy Bradford No. 3

rious independent oil company. Although only in his twenties, Don was well-versed in drilling and sizing up an operation. He spotted that the drill rig had been shut down for a week and Ed could not have been taking core samples as bragged in the telegrams. Unlike his oil scout counterparts, Don was extremely likeable and his heart bled for the driller struggling every day with a run-down rig. His admiration for Ed soared when he learned that the well was bottomed at 3,400 feet — 900 feet deeper than the rig was designed.

When Reese asked Ed for cuttings for examination by a paleontologist to determine the formation where the well was bottomed, Ed consented, surprising the Sinclair scout. Drillers never give away a well's secrets unless bribed. To make sure the cuttings weren't salted, Reese sneaked back that night to snitch more cuttings. Reese was unaware that Ed also wanted a professional opinion. Doc Lloyd, the pseudo geologist, had not checked the cuttings or inspected the well in two years. The results were devastating. The geologist's opinion was that the well had not reached the Austin Chalk, the Woodbine was another 1,500 to 1,700 feet deeper and, at that depth, the Woodbine would not contain oil evolved from tiny organisms trapped in the prehistoric seabed, but salt water from an ancient ocean.

Another friendly oil scout (all good oil scouts must be friendly), was Hank Conway from Amerada Oil, another large independent oil company. Hank admired Ed and believed the driller was wasting his time trying to bring in a well in the unproven barrenness of East Texas. He also knew that Ed wasn't taking core samples as his telegrams stated, so he volunteered to do the unthinkable — furnish Ed with a core barrel and have the sample examined by a paleontologist to determine the truth.

It was a gamble Ed couldn't refuse. He was at 3,456 feet, 94 feet from where Doc Lloyd said the well would hit the Woodbine. Ed knew Doc was a fraud, but that's all he had to go on. Amerada's report on the core analysis were similar to the results by Don Reese's geologists...maybe worse. They believed the Woodbine would not be found in the area but, if it was hit it would be at 5,100 feet and contain the remnants of the prehistoric sea — salt water.

Ed knew he couldn't baby the old drill rig down to 5,100 feet. He decided to drill to Doc's target for the Woodbine, 3,550 feet, even though geologists advise that the best geological investigation produces mere estimates of the depth of a formation. Also, Doc's original site selection for the well was almost two miles to the west.

Several days later, Ed noticed that the drill was whipsawing, a sign the bit was cutting a hard formation. When he examined the cuttings, he discovered sand streaked with an oily rainbow film — "color" in the vernacular of oilmen. He ordered the crew take a core sample, lay it on the rig floor then go home. If it was oil, he didn't want to raise their hopes or have them celebrate in town and blab about it when they had a few drinks too many. He decided to put the core barrel in the trunk of his car and examine it when he arrived home.

That evening behind a garage, Ed Laster saw the most beautiful sight he had ever seen. A core of the Woodbine sand dripping oil.

WOULD YOU TRUST A SINCLAIR OIL SCOUT?

Sinclair Oil's president and founder, Harry Sinclair, was Colonel Stewart's co-conspirator in defrauding their corporate shareholders of $8.3 million. He had been convicted of contempt of Congress and jury tampering and served time in jail in connection with the bribery of Secretary of the Interior Albert B. Fall in order to obtain an oil lease on the U.S. Navy's Teapot Dome oil reserve. Sinclair was ordered to pay the Navy $12 million for oil illegally removed from the Navy reserve.

However, Harry Sinclair's illegal activities did not prevent him from being unanimously reelected to the board of directors of the American Petroleum Institute, Big Oil's powerful lobbying organization in Washington.

8
DO UNTO OTHERS

The teachings of Moses and Jesus evolved into the *Golden Rule: Do unto other as you would have them do unto you.* In the rough and tumble oil patch, there was a variation of the *Golden Rule: Do unto others before they do it unto you.*

Ed's exhilaration at finding the Woodbine turned to panic. He remembered leaving the bucket of cuttings at the rig. If an oil scout found the Woodbine cuttings, the news would break and the price of oil leases in the area would skyrocket. Columbus could not afford to buy leases at the inflated prices for himself or the leases he owed Ed as part of his wages.

His panic turned into dismay when he arrived back at the drill rig. Someone had taken the bucket of cuttings!

Dejected, the driller sat on the derrick floor staring into the black night that had begun as a dream of riches then twisted into a nightmare. How could he break the news to Columbus that he had hit the Woodbine, but an unknown oil scout had discovered the bucket of cuttings? Also, last week, Ed had sold a small lease a few miles from the well because Columbus had not paid his wages. No doubt, the lease would be worth thousands more than he sold it for. His bad dream came true when the Sinclair oil scout, Don Reese, appeared laughing. Don had taken the bucket of cuttings and hidden it.

Don merely chided him for salting the well and told him that he wouldn't tell anyone. Ed had no recourse but to laugh and tell Don to believe anything he wanted, but suggested his friend pick up a few leases in the area.

On his drive back to Henderson, Ed's mind bolted like a mustang. There were thousands of dollars to be made. Two years of poor boying and times without pay were over, *if* Columbus could raise the money for the leases he had promised and many thousands more, *if* the well came in a gusher.

As the say in Texas, Ed was happy as a pig in slop when he telephoned the old promoter and told him: "We hit the Woodbine! I have nine inches of oil sands."

"Oh," Columbus said casually. "Put a sample on the bus and be quiet about it."

Dumbfounded at Columbus' apparent lack of interest and failure to mention the 360 acres in leases he owed him, Ed listened silently.

"I'll be there in a day or two with Doc Lloyd," Columbus said. "From now on Doc's in charge of the well."

Ed slammed down the receiver and stared at the phone. "You bent over, old, crooked son of a bitch, you ain't gonna cheat me!"

◆ ◆ ◆

The next day Ed showed the Houston manager of the Mid-Kansas Oil & Gas Company a piece of the core sample on the promise of a 1/4 interest in a 1,100 acre lease near Daisy's farm Mid-Kansas leased at $1.00 an acre.

It was hate at first sight. Doc appeared with a driller and advised Ed he was being replaced, but would work nights under Doc's direction. Ed objected to Doc's plan to cement the casing without reaming the hole, as the bottom 1,800 feet had been drilled with a six and one-half inch bit, unlike the eight inch bit on top. Further, Ed argued, a proper drill stem test could not be made without endangering the rough hole.

Ed and the crew listened in stunned silence as Doc explained a new technique for completing a well. But when he finished, a mule skinner would have blushed at Ed's language.

Don Reese was in a quandary. He liked Ed Laster and had difficulty believing he would salt a well. His search of the Rusk County land records had failed to show that Columbus or Ed had recently recorded buying new oil leases. However, that wasn't a true test, sharp promoters often held back recording leases until they completed a large block for fear the bonus price would increase. The telling information found in the records was that Ed had sold a lease last week. No driller who thinks he's close to hitting pay sells his leases.

A few days later Don stopped by the Daisy Bradford No. 3 and was surprised to see Columbus, Daisy and the Tuckers. Ed and Doc were pulling another core sample. Don was met with unfriendly stares from all but the ultimate con man.

"Mr. Joiner, I need a piece of that core," Don said, "or I could be in real trouble with my company."

Columbus winked and pinched off a thumb-sized piece of soft sand from the core and handed to him . "Can't let that happen," Columbus said. "Thanks for all your help."

◆ ◆ ◆

That night the oil scout sneaked back to the well, but could find no evidence of the Woodbine oil sands. The crew had scrubbed the rig floor.

The following morning Don was the laughing stock of Sinclair Oil. He should have known that Columbus Joiner salted the well so he could fleece more money from widows.

Ed found a surprising ally when Daisy called and asked him to meet her and her brother, Clarence Miller, at his Just Right Motel. Daisy didn't think much of Doc's idea either. The flinty-eyed widow announced: "I'm staying with the breeze that carried me this far."

Daisy spit fire. Columbus' lease ran out the following day and it was her turn to play hardball. She was not going to renew the lease unless Ed was in charge of the drilling, Columbus furnished satisfactory equipment to complete the well — no more third-hand junk — and Columbus assigned 100 acres of her lease to her brother.

The next evening the old con man realized he had no alternative but to sign the agreement Daisy demanded. The cocks that had crowed he was more interested in selling leases and stock had come home to roost. Everyone had ganged up on him. They knew the well was valuable and were taking advantage of his poor financial position. Daisy was getting even for his late rental payments, failure to produce a well for five years and flimflamming her out of one-quarter of the standard royalty.

"You're working against me," Columbus told Ed.

Ed replied: "Not if you really want a well."

Ed had another mood swing — from elation at back being the driller on *his* well to outrage and fear: Doc had drilled 75 feet deeper and through the Woodbine!

According to all oilmen, from experienced petroleum engineers and drillers down to the greenest weevil, Doc Lloyd's drilling was extremely dangerous, and risked drilling into salt water generally found beneath oil and gas. Salt water could have flooded and destroyed the well.

Ed raced to Overton to face Doc, but the eminent geologist, petroleum engineer and scoundrel had already skipped town.

AN UNANSWERED QUESTION

There has always been a lingering question whether Columbus and Doc planned to sabotage the well. There was no reason to drill 75 feet through the oil-bearing Woodbine. As years go by, Columbus "Dad" Joiner's legend grows and he becomes more revered and it is rarely mentioned. Read on and come to your own conclusion.

The Sermon on the Mount was taught in all Sunday schools in fundamentalist East Texas. Walter Tucker and R.A. Motley knew the gospel according to Matthew: Chapter 7, Verse 12: *Therefore all things whatsoever ye would that men should do to you, do ye even so to them.* When the two gentlemen learned that Columbus had failed to transfer leases to Ed Laster, they helped the driller acquire oil leases covering 200 acres.

Columbus wheedled a free drill-stem test from the Miller Brothers of El Dorado, Arkansas, who had developed a new drill-stem method to test the pressure and amount of fluid in a producing formation. Aware Columbus had a poor boy well and was infamous for never having sufficient funds to drill and complete a well, the Millers invited a friend to accompany them. If the drill-stem test proved successful, the friend would offer to complete the well for an interest in the production. The Millers' friend was H.L. Hunt.

H.L.'s accomplishments were few the last three years, although keeping two wives happy and unaware of the other is quite an achievement. Frania presenting him with a son, William Herbert, and Frania pregnant with Helen may be considered accomplishments. The biggest changes in the former gambler's life were he had moved Frania and her children to Dallas and he was devoting most of his time to the oil business, which he likened to gambling.

For all H.L.'s bragging about his successes as a poker player, he seldom gambled in the oil business. Wildcatting was not his style. The odds of hitting a wildcat well were worse than filling an inside straight. Moving into an area with his oil scouts and buying leases after a wildcatter took the risks was more to his liking.

Folks from Overton and Henderson, anxious to see if Columbus' ocean of oil was true, and a silent group of oil scouts and geologists, wondering if they were wrong or the drill-stem test was a scam, made up the crowd when H.L. stepped from his car. All eyes were on the crew lowering the tool into the hole in search of the elusive black gold hidden in the rocks for millions of years.

H.L. carefully folded his jacket and placed it in the front seat of his car and adjusted his straw hat as he sized up the spectators, concentrating on the oil scouts from Shell, Amerada, Texaco, Sinclair and Humble...the Big Oil competition in the race for leases.

One of the Miller brothers saw his poker-playing friend light an eight-inch stogie and work his way to the front of the crowd. His introduction was simple: "C.M. Joiner, meet H.L. Hunt."

Eyeing each other like a pair of tomcats, they shook hands and marked the beginning of their friendly double-dealing in which Columbus would always call H.L. "Boy." No doubt, H.L. began their conversation with the standard opening line of all oilmen: "How deep is she?" Oil wells are female. It is a form of phallicism.[8] Oilmen also inevitably bring the well's geology into the conversation. H.L. was an

---

8 My driller's pickup truck in Oklahoma had a bumper strip that read: *DRILLERS DO IT DEEPER.*

avowed "creekologist" in his early years in the oil business and bragged that Hunt Oil Company did not have a geologist on its payroll until after World War II. Of course, if you had oil scouts (spies), you didn't need geologists.

H.L. was disappointed when Columbus told him that D.H. "Dry Hole" Byrd was financing the completion of the well. He would have to be satisfied with being an onlooker and act fast if the old con man's drill-stem test proved positive.

CREEKOLOGY

Creekology evolved as a pseudo theory that oil is found near creeks after the first discovery of oil in Pennsylvania, the most famous being Oil Creek. The discovery of the world's largest oil fields in Saudi Arabia's desert should have debunked the fiction, but my Oklahoma driller still swore by it in 1984, and I could never prove him wrong...Oklahoma has creeks most everywhere.

The odor of gas filled the air after the drill-stem was lowered.[9] Quickly, the crew started pulling pipe from the hole. Before they could complete the task, the rig began to vibrate and a deep rumble from below turned into a roar that could be heard for a quarter of a mile. Mud, gas and oil shot to the top of the derrick and one of the derrick's legs snapped and splintered with a loud crack. It was over as quickly as it began. Nature's furious blast of hydrocarbons trapped beneath the surface for millions of years had erupted in a gigantic belch.

---

9 The odor of natural gas was not what you detect from a gas leak in your home. Natural gas is odorless and requires an additive for safety reasons. Gas from a well often contains poisonous hydrogen sulfide, which smells like rotten eggs and must be removed. As far as oilmen are concerned, it smells like money — beauty is in the nose of the beholder.

A tide of well-wishers (no pun intended) rolled towards Columbus. The old man gave them a wizened smile, unable to accept their congratulations. It wasn't an oil well yet.

Ed remarked that it was going to be a pretty good well, "if we can bring it in."

H.L. said nothing. He was too busy thinking how he was going to acquire leases.

Later, the most reproduced photograph in the history of the petroleum industry was taken. One story is that R.A. Motley lined up Columbus, Doc, Ed and the drilling crew in front of the derrick. Glenn Pool stood behind Columbus and Doc shaking hands. Ed changed into a white shirt for the occasion. (Dennis May was probably off someplace enjoying a snort of corn liquor.)

A moment before the picture was snapped, Columbus motioned to H.L. "This picture's going down in history, Boy, you'd better get in it."

H.L. posed for the picture wearing a straw hat and chomping on a cigar. Columbus was right, the photo would go down in history, but he didn't know how big a part H.L. Hunt was going to play in the history of the East texas oil field.

PART TWO

THE BOOM

Dear Lord: Please send us another oil boom. We promise not to screw this one up.

Bumper strip seen in Texas after the oil boom collapse in 1986. It could also have applied to the 1930's boom.

9
THE BOOMERS

The night of the drill-stem test, H.L. met with two Hunt Oil Co. men, Jick Justice and Charlie Harden, experienced drillers and oil scouts. Jick, a former mule skinner, carried the reputation of being mean as hell, as did the heavy drinking Charlie. Most important, they would do anything for their boss and kept their mouths shut.

H.L. had to act fast. Independent oilmen don't require a producing well to start a boom. The smell of oil attracts them like flies to horse manure. The price of leases had been known to jump a hundredfold overnight on the rumor of an oil strike. Jick and Charlie were instructed to lease every available acre in the area of Daisy Bradford's farm.

The first thing H.L. arranged was rooms for himself and his men in the Just Right Hotel, aware there wouldn't be a hotel room for fifty miles when the news broke. The next order of business was keeping an eye on Columbus Joiner.

JOINER FINDS OIL ON MILLER FARM read the banner headline on the *Henderson Times*. It was the biggest news in Texas since Sam Houston kicked Santa Anna's butt at San Jacinto in 1836.

There was no way to stem the tide. First they arrived from neighboring counties, then Arkansas and Louisiana. Then the forerunners of the small farmers portrayed in Steinbeck's *Grapes of Wrath* rolled in from Oklahoma. Cities, such as Detroit, sent thousands of auto workers Henry Ford had laid off.[10] It was the first glimmer of light in the

---

10 Henry Ford would have enjoyed this novel history. He was quoted: "History is more or less bunk."

darkness of the nationwide Depression. East Texas, which had struggled through poverty and drought for a decade, was about to leap from privation overnight. It was more than a dream of from rags to riches or a quick buck to be made. It meant jobs, just like Walter Tucker said it would.

To the independent wildcatters, it meant little that Columbus Joiner's drill-stem only showed a trace of oil. They only wanted a chance to drill. The major oil companies, always cautious, sent more oil scouts (spies).

Petroleum Geology 101 1/2: Evidence of oil is found in numerous locations in the earth's mantle, but that doesn't mean you can get it out. Millions of years ago, Mother Nature trapped billions of minute organisms at the bottoms of ancient seabeds, reefs and deltas that eventually evolved into the hydrocarbons we call oil and gas, and she is not willing to give them up without a struggle.[11] That's why Ed Laster cautioned: *If we can bring it in.* Oilmen are concerned about *porosity* and *permeability.*

Porosity refers to the space between the particles of sand or sediment in the ancient seabeds available to contain oil or gas...and hopefully not salt water. Permeability refers to the connections between the spaces. A porosity of 10% is generally necessary in a producing field. However, even 10% can be difficult to find. The weight of thousands of feet of rock on top of the sand squashed the hell out everything over tens of millions of years, which is one of the reasons oilmen often can't coax 50% of the oil in a pool to the surface.

Permeability is necessary for the oil to migrate between the grains of sand to the well.[12] If the sediment is not permeable, the oil cannot be drawn between the grains, which is another reason we cannot extract all the oil. Also, if the oil or gas is not trapped in a sedimentary formation by an impervious cap rock, in all likelihood any oil or gas that was there has migrated to another location.

---

11 Contrary to old Sinclair gasoline station signs and ads, oil is not the slimy remnants of dinosaurs, although a brontosaurus or two may have contributed a few barrels.

12 Permeability is measured in millidarcies, named after a guy named Darcy who knew a lot about hydraulic conductivity. Few oilmen understand millidarcies. They merely ask a petroleum engineer: "Is the permeability good?" Of course, if the oil gushes out, they may not bother asking.

Geologists will tell you that oil is *fugacious,* which comes from the same Latin derivation as the word *fugitive* — it tries to escape. Oil under your land can migrate to your neighbor's land. As a result, oilmen hurry to drill wells before the oil migrates (escapes) from their leases and to encourage oil from the neighboring land to migrate (flee) onto their land before the neighbor can get his fair share. Appropriately, it is called the *Rule of Capture.*

The foregoing brief lesson from Petroleum Geology 101 1/2 is important for readers to understand when all hell breaks loose and oilmen battle each other and the government over oil in East Texas.

Permeability and Porosity mean little if there is not enough oil in a pool to make it worth going after. On occasion, a well will roar in gushing and turn into a trickle. This often separates the major oil companies from the small independents. Big Oil's corporate bureaucracy is concerned about paying dividends every quarter. How would you like to have stood in the shoes of the president of Standard Oil of Ohio (Sohio) before your stockholders in 1983 and announce that you had just drilled a wildcat well in Mukluk, Alaska, that cost over $2 billion ("billion" — not a misprint) and struck salt water? Small independents, although now a dying breed, merely have to sniff oil, hear a rumor or have a hunch. As a result, 80% of the oil discovered onshore in the United States has been found by independents.

The majors and larger independents had written off East Texas before the drill-stem test. Their bloated bureaucracies were still not willing to believe it. Most Big Oil geologists believed it was a small pool stumbled upon by a con man and not worth their effort or were too embarrassed to admit they were wrong. The exceptions were Sinclair, Shell and Humble.

HUMBLE OIL AIN'T HUMBLE

Humble Oil & Refining Co. didn't get its name because it was modest and unpretentious. It was incorporated in 1917 by independent oilmen and named after Humble, Texas, a small town a few miles north of

Houston, to compete with Gulf and Texaco. Two years later, it crawled into bed with despised Standard Oil of New Jersey (Exxon) when it sold 50% of it's stock to the predator. It marked the first crack in the Texas antitrust laws barring foreign (out of state) corporations from doing business in Texas. By 1925 Humble was the largest oil producer in Texas, and by 1928, the largest pipeline company in the United States. It's control over prices was evidenced by purchasing three times more crude oil than it required for Standard Oil's refineries. It exported the rest. (Remember the Achnacarry Agreement and price fixing in bonnie Scotland?)

Humble's president between 1922 and 1932, William Farish, was so ruthless, he was elected president of Standard Oil of New Jersey. There are those who claimed, not without reason, that Farish cooked up the deal to merge with Standard with Walter Teagle so Standard could sneak back into Texas. Farish and Teagle had been pals since they worked together on the National Petroleum War Service Committee during Word War I.

Henderson and Overton couldn't hold the thousands rushing to the oil boom. The Just Right Hotel crammed four and five to a room and tripled the rates. Food prices skyrocketed. Hamburgers selling for five cents before the burp of oil from the Daisy Bradford No. 3 now cost a quarter.

Wildcatters, lease hounds and roughnecks weren't the only ones rushing to Rusk County. Equipment salesmen, mechanics, machinists, carpenters and many who just needed a job — any job — to feed their families poured into the area. With them came the oil patch trash — bootleggers, gamblers, con men, pickpockets and prostitutes. The hookers were picturesque in uniforms of bright red, yellow and blue beach pajamas as they strolled brazenly along the streets twirling parasols and selling their wares for $2.00.

Overnight a town christened "Joinerville" rose along the highway seven miles west of Henderson where the one lane dirt road turned off to Daisy Bradford's farm. Joinerville was well-named. The poor boy shantytown was a slipshod collection of shacks and tents advertising hotels (*$3 A NIGHT — 2 TO A BED*) and saloons (*5¢ BEER*). Crude hand-lettered signs read: *HAIRCUT $1 — BATH $2...HONEST POKER INSIDE...GIRLS DIRECTLY FROM NEW ORLEANS...TOILET 10¢*. The ramshackle town was double the combined populations of Overton and Henderson within two weeks and would have held many more, if it had plumbing.

Rusk County farmers enjoyed the bonanza. The price of oil leases hit $400 an acre — ten times what the barren land would have sold for weeks earlier, if you could have found a buyer. Within two weeks of the drill-stem test, over 2,000 land sales and leases were recorded in the county courthouse. Some of the shrewd farmers held back leasing all their land. If they could get $400 an acre now, what could they get *if* Joiner really strikes oil?

H.L. acquired leases on 400 acres to the east and south of the Daisy Bradford No. 3 with the little cash he had in his pocket when he arrived, earned playing poker and betting on the Philadelphia Athletics in the World Series. While his heart was with the St. Louis Cardinals, his wallet told him to bet on the 1929 World Series champs and their brilliant manager, Connie Mack, and pitcher, Lefty Grove. H.L. pocketed $2,500 at the end of the six game series.

Not all of Big Oil was asleep. Humble oil gobbled up 16,000 acres in the area, which would prove to be 13% of the oil field. Acting fast, it's average lease cost was only $20 an acre for the first 12,000 acres. Eventually, it would own 16% of the Black Giant.[13] Shell and Sinclair also quietly picked up oil leases based on tips from Ed Laster when he was pissed off at Columbus.

---

[13] In its self-serving "autobiography," *History of Humble Oil and Refining Company: A Study in Industrial Growth*, it claimed it would have discovered the Black Giant six months earlier, but hadn't gotten around to drilling the well. ®

MINERAL RIGHTS

In Texas and other mineral producing states, the ownership of the land's surface is often sold separately from the mineral rights. The split estate empowers the mineral estate owner or his lessee to use the surface and do anything reasonably necessary to extract the minerals. This can cause a hell of a brouhaha, if a mineral rights owner leases the land and a farmer, who only owns the surface rights, wakes up one morning and discovers a drill rig tearing up his alfalfa field.

Oilmen need maps to determine the ownership of the mineral rights — not the type of maps service stations used to hand out free years ago when you filled up your tank, but plats diagraming the property boundaries and the name of the mineral owners. In many booms, land is bought, sold and leased so often, the county clerks are unable to keep up with the transactions and the records do not show where the gushers and dry holes have been drilled.

Enter W.W. Zingery from Fort Worth, an oil field map maker. There are two opinions of Zingery. Big Oil industry flacks claimed he was a dedicated map maker with the best interests of legitimate oilmen at heart. Most East Texans said he was a no count, sneaky son of a bitch.

In order to get a head start on his map making competition, the first night Zingery hit Henderson, he and a couple of independent oilmen pals, visited a Humble *Landman*, who had men mapping part of the area. While his two cohorts were plying the Humble landman with some of the finest corn liquor in East Texas in one room, Zingery traced the Humble map in the kitchen. No doubt the independent oilmen didn't think they were doing anything wrong, the landman worked for Humble, a subsidiary of the Standard Oil of New Jersey, the nastiest and biggest member of Big Oil. The following day, Zingery was at the courthouse updating his map. When he couldn't finish by the end of the day, he slipped the clerk a few bucks to let him slip in after closing in order to beat his competition, proving Zingery was a sneaky SOB.

LANDMEN

A landman's job is to obtain mineral leases from landowners, including negotiating the lease bonus, royalties and terms of the lease. Oil companies often hire independent landmen when they do not want other companies to be aware they are interested in the area or the landowner to know who he is dealing with. Oil and gas leases are freely assigned. It is often a smart move to hire an independent landman. Today, if a landman admits he works for Exxon, even the dumbest hick in the backwoods knows that Exxon is loaded and will jack up the price. Also, since the Exxon Valdez spilled 260,000 barrels of crude oil in Alaska's Prince William Sound, there are many environmentally concerned citizens who might tar and feather the Exxon landman.

Many landmen belong to American Association of Petroleum Landmen, which later adopted a badly needed code of ethics. Although women are now engaged in the field, don't expect the good ol' boys to agree to call themselves *Landpersons*.

Within days, Zingery was raking in thousands of dollars a day selling updated maps and, like everyone else, wheelin' and dealin'. It was inevitable that Zingery would pick up several $100 Rusk County Oil Syndicate certificates as payment. It was also inevitable that Zingery would run into another type of character in the courthouse who preys on people — sneakier that an oil scout, more devious than a used oil equipment dealer and with less morals than the hookers parading outside in beach pajamas...*a lawyer.*[14]

Zingery and the lawyer were as smart as Dea England. They noted that all 300 certificates entitled the bearer to four acres out of a 320 acre lease parcel. Even the lawyer could multiply 4 X 300 and calculate that Columbus should have set aside 1,200 acres. As a typical lawyer, he wanted to make a "federal case" out of everything. The lawyer told Zingery he planned to file suit against Columbus on behalf of a widow, who owned several certificates, in the federal court in Dallas, where he and his fellow big city lawyers could flimflam the country folk easier.

Fortunately, lawyers are prone to bragging. (In those days, it was considered unethical for attorneys to pitch their services in the yellow pages between *astrologers* and *auctioneers*.) Alarmed, Zingery contacted a local lawyer. No country lawyer wants to lose the hometown advantage. It would also keep the legal fees in East Texas. The local lawyer called Judge R.T. Brown at home. Getting a Judge to come to court on a Saturday is part of the hometown advantage. The fact that Judge Brown was later called the "Sage of East Texas" proved that. Before the judge returned home to milk the cows, the involuntary receivership in the case of *W.W. Zingery v. C.M. Joiner* was filed.

---

14 The author regrets having to admit he ~~is~~ was a lawyer. Letters of indignation from lawyers must be double-spaced in 12 point bold type (no italics) on letter-size paper or they will be returned unread to the pompous pettifoggers. Documents over two pages, containing Latin words or including footnotes will be contributed unread to the Rusk County, Texas, paper recycling program. All correspondence from attorneys must be accompanied by a non-refundable $1,000 retainer to be held in trust for their unwed mothers, if they can prove who their father is.

Columbus' oil leases and the Rusk County Oil Syndicate certificates were tied up in the East Texas court and scheduled for a hearing in thirty days.

TEXAS JUSTICE

A lawyer who knows the judge is better than a lawyer who knows the law.

10
GUSHER

East Texans were shocked. The man revered as *The Daddy of the Rusk County Oil Field* was accused of overselling the third Rusk County Oil Syndicate by 350%. Investors holding certificates from the first two syndicates claimed they should share in the third syndicate because Columbus failed to complete the first two wells. Copies of oil leases he had sold over the years, when finally filed with the county clerk, revealed he had peddled them to the unsuspecting over and over and over. One lease had been sold eleven times.

It was true. Columbus knew it. Dea Knew it. Doc knew it. The people of East Texas couldn't deny it, but that didn't bother them. Columbus Marion Joiner was their redeemer who delivered them from perdition. Who couldn't help loving the old man? He was the "Daddy" of their oil field and forever would carry the name "Dad Joiner." (As everyone else started calling him "Dad," so will I.)

Rumors that "Big Oil" was trying to steal the oil field from the "little guy" spread like burning oil, with Doc Lloyd in the forefront of the campaign. Dad refused to defend himself. He didn't have to. The farmers and townspeople were behind him. Their opinion was reflected in an editorial in the Tyler *Courier-Times:*

> Is [Dad Joiner] to be the second Moses to be led to the Promised Land, permitted to gaze upon its milk and honey, and then denied the privilege of entering by a crowd of slick lawyers who sat back in their palatial offices cooling their heels and waiting while old "Dad" worked in the slime, muck, and mire of slush pits and sweated blood over his antiquated rig, down

in the pines near Henderson? Is it right to tie this poor fellow up in court while the horde of "big boys," which has flocked into the territory adjacent to his holdings, sink holes and sucks the oil which he discovered, right out from underneath him?

We are sure that we voice the sentiment of the masses when we say emphatically — No!

If they are permitted to tangle him up in court, we predict that other men of small means, land owners and farmers, will follow him to the guillotine of financial execution. East Texas has had enough of inquisitious "oil trusts" without sitting quietly by any longer while the "big boys" crowd in, leaving "Dad" Joiner out of the picture ...

It is high time the independent oil operators had their inning.

And, lastly, if they haul "Dad" Joiner into court, you farmers and small landowners should turn him back his oil well.

Now, if this be bolshevism, then we're bolshevists.

Dad Joiner also had two unsung heros in his corner, R.A. Motley and Walter Tucker. Little did it matter he was a con man, Dad had brought prosperity to Rusk County. Nor did they care that Dad's oil well had yet to produce the first barrel of oil. They decided it was time to celebrate and honor the Daddy of the Rusk County Oil Field. Walter went to work to assure it would be the biggest party East Texas had ever seen.

Overton's "Joiner Jubilee" was a wingding. (It couldn't be held in Henderson, the county seat. Motley's bank was in Overton.) Banners stretched across Overton's dirt streets and Old Glory and Lone Star flag draped every building. Ten times the population of Overton watched the Boy Scouts, American Legion, fire departments and high school bands march along the street. A float representing the Daisy Bradford No. 3 and cars containing its faithful crew drew the biggest applause.

No one actually recalls the special lyrics glee clubs sang to familiar tunes, although one tale is that the Methodist Men's club repeated the refrain: *We'll hang old man Rockefeller to a sour apple tree...we'll hang old man Rockefeller to a sour apple tree...*☺ They were followed by the Baptists, which everyone knows can out sing the Methodists, singing to the tune of *God Bless America:*

God bless Dad Joiner, man that we love,
Stand beside him and guide him through the fight with a light from above.
From Overton, to Henderson, through Rusk County, save our oil,
God Bless Dad Joiner and reward him for his toil. ® [15]

On a platform in front of the Overton State bank they honored Dad as a hero. Streamers unfurling overhead, he straightened his crippled torso and accepted the accolades and love the people bestowed from their hearts. As he stood returning the waives and cheers, his face beamed and tears glistened from the sight and sound of the adoration. That afternoon and into the night the celebrants danced and feasted on barbecued beef and pork. Everyone had to shake Dad's hand. One has to wonder what was running through his mind. Next week, would they call him a swindler and condemn him to drown in his ocean of oil?

The boom continued into the following week as Ed Laster and the crew babied the well in preparation for bringing it in. Dry Hole Byrd was true to his word and supplied new equipment and tanks to hold the ocean of oil. Word spread that on Friday Dad was going to complete the well. It might as well have been declared a holiday, few went to work. By morning the crowd had swelled to 10,000. A carnival atmosphere swept Daisy's farm, clogging the roads with cars and buggies for miles. They laughed, joked and drank everything from Coca Cola to corn liquor. It mattered little that the crew's task of cementing the casing was dull and tedious, they had come to see the birth of *their oil well.*

To the crowds dismay, Dad did not appear. His friends said he was ill, but knew he was avoiding meeting his investors. However, Dad could not stay away the next morning. Ed and the crew were scheduled to bail and swab the well to clear it of water and mud, and he had to be there to see his well come in. Everyone was disappointed when night fell and the exhausted crew shut down the rig.

---

15 I don't believe it, but the lines were too wonderfully hokey not to include. One old-timer recalled seeing a mule draped with a banner: *STANDARD OIL IS AN ASS.*☺

October 5, 1930, was Sunday in fundamentalist East Texas, but there was never a question if the crew was going to work. Every church in the county prayed for a gusher. Nevertheless, three days of strain began to take its toll and doubt lines creased the faces of the fainthearted. D.H. Byrd lifted their sagging spirits by telling of his misadventures and that D.H. stood for "Dry Hole." The wildcatter had drilled over fifty dry holes before hitting his first gusher...then brought in a second gusher the same day.

Late in the evening, Ed climbed down from the rig and announced he had to shut down because they had run out of firewood for the boilers. Within minutes Motley and Tucker had farmers removing the spare tires from their trucks and cars and stoking the boilers with strips of rubber until the black smoke and fetor of burning rubber hovered over the thousands of faithful onlookers.

Daisy, her fingers crossed, was sitting with the Tuckers at the table where Leota had served the crew meals when she noticed nightfall would soon be upon them. As the crew withdrew the swab from the well for what she thought was the thousandth time, she felt a vibration under her and heard Ed shout to shut down the boilers and put out cigarettes.

The crowd stared at their feet. Tremors buckled their knees. A distant rumble turned into hell's thunder and the derrick began the shake and rattle. Pandemonium stuck when the crew leaped from the rig. With the roar of a locomotive, a black pillar gushed high over the vibrating derrick and rained on the throng.

Cheering, they turned their faces up to God and showered in the black slime...the salvation of Rusk County. Grins widened with each splatter of oil on their faces and anointing their heads. Others caught it in their hats or rolled on the oily grass. Dennis fired his pistol into the black cloud until it was wrestled from him for fear it would ignite the billowing gas. Daisy danced in the black rain. The quiet ones, like Walter Tucker, celebrated by thanking God for answering their prayers.

Drenched in oil, Ed and the crew turned the valves and the gusher slowly dropped, like a fire hose running out of water, and shunted the oil into the storage tanks.

Dry Hole Byrd read the tank gauge and whispered to Dad, allowing him the honor of telling the crowd: *Sixty-eight hundred barrels a day. Dad Joiner had brought in a gusher and made good on his promise of an ocean of oil.*

Gusher

♦ ♦ ♦

"I dreamed it would happen, but I never really believed it," Dad whispered to Doc as they faced the cheering crowd.

♦ ♦ ♦

Malcolm Crim grit his corncob between his teeth. The gypsy had told him there was oil under the neighboring farms ten years ago. Now he had to convince someone that his farm fourteen miles to the north had oil beneath it, too.

♦ ♦ ♦

Squinting at the gauges, H.L. Hunt whispered to his oil scout, Charlie Hardin, "Looks to me more like sixty-six hundred barrels a day. But with oil at a dollar fifteen cents a barrel, I'd be happy with seven thousand five hundred and ninety dollars a day."☺

Any man who could cipher as fast as H.L. was a man to be reckoned with.

11
WHEELING & DEALING

Years of drought ended. The Lord blessed the farmers with rain. The first day the cloudbursts ran off the arid land baked by the hot summer sun. Then the Lord cursed them with too much rain. Was it an omen? Creeks swelled and roads and fields turned into an ocean of mud. Cars and trucks sank up to their axles in the red Texas clay.

As the downpour lashed East Texas, major oil companies and their geologists believed they had the last laugh. The flow of oil from the Daisy Bradford No. 3 became erratic and dwindled to 250 barrels a day. In oilmen lingo, it flowed in heads, belching a 100 barrels, then gasped and died for hours.

The life of an oilman is not for the faint of heart. When not keeping his eye on Dad Joiner's irascible oil well, H.L. was in his hotel room plotting and directing his oil scouts. He concluded the Daisy Bradford No. 3 had hit the upper edge of a pool and was being fed erratically. Studying a map of the dry holes to the east, he surmised the oil field lay to the west, south or north. To his chagrin, three of the four leases he had acquired were in the east and only one in the south. He sent for his antiquated drill rig in El Dorado, described as a hair better than the rig Dad had used on Daisy's lease, to drill on the lease to the south.

Desperate for cash, Dad sold the oil production from Daisy's lease for drill rig boiler fuel with little or no regard to its real value. The crude oil was over 37° API gravity and sweet. Light oil was highly desired by refiners because it produced greater quantities of gasoline and contained little sulfur. H.L. contacted Sinclair Oil & Refining in Houston, which agreed to build a loading rack on a branch line of the

Missouri-Pacific Railroad, if he built an oil gathering pipeline from the field to railroad. The pipeline would assure H.L a market and method of transporting his crude oil and give Sinclair access to the field. Neither would have to deal with Humble Oil, known for attempting to control the pipeline transportation and, thus, the price of crude oil. The gathering line would also earn H.L. 15¢ a barrel for transporting other operator's oil.

Charlie Hardin, H.L.'s savvy oil scout, reported that while Humble's geologists were still denying there was oil in the area, Humble landmen were buying leases.

"That's why I don't need college boy geologists," H.L. said. "Frank Foster is drilling a well for the Deep Rock Oil Co. a half a mile west of the Bradford lease, right in the middle of Joiner's 5,000 acre block of leases. I want a man to sit on the well and report every detail to you. I promised Frank a little cash if we're the first ones to know if the well hits oil."

Charlie's report that there were no leases available near the Daisy Bradford No. 3 was brushed aside by H.L. "There's Joiner's leases."

No one knew where Doc hid from two wives, who saw his picture in the newspapers standing in front of the Daisy Bradford No. 3, as they searched Overton looking for him. Only Dad's closest friends knew where he was hiding in Dallas. The pressure of Zingery's lawsuit and the cold rain had laden him with remorse and the flu. One can only wonder what thoughts of self-pity tormented him. He was facing a lawsuit that could strip him of his honor as well as his ocean of oil. Two month's earlier, he had sold an eight acre lease for $125 in order to pay the crews wages. Last week, four acres of the lease were sold for $10,000 and a 1/16 overriding royalty.

When the day of reckoning arrived, Judge R.T. Brown's firm jaw and steely eyes dominated the courtroom in Henderson. Everyone involved in the lawsuit appeared but the defendant. The plaintiffs who groused when Dad's lawyer, Colonel Bob Jones, advised the judge that Dad was too ill to attend the hearing were shouted down by the true believers. Judge Brown's tolerance lasted but a few seconds. The banging of the gavel brought a hush over the crowd and the Judge ordered the attorneys to briefly state their case.

Wheeling & Dealing

When Colonel Jones concluded his final argument, Judge R.T. Brown, the Sage of East Texas, issued the must famous ruling in the annals of the Texas judiciary:

> *I believe that when it takes a man three and one-half years to find a baby, he ought to be able to rock it for a while. This hearing is postponed indefinitely.*

Zingery had his fill of East Texas justice and withdrew his lawsuit. Needless to say, the map maker was not popular with the locals. Men outside the courthouse patting their guns and mumbling that Zingery would have been a lot safer in the Alamo than he would be in East Texas helped him make up his mind.

A few weeks later, another case was filed in Dallas; however, Dad was nowhere to be found to serve the complaint. His usual haunt, the Adolphus Hotel, denied he was registered. However, a sharp lawyer's $100 tip to a bellboy unmasked Dad's room where he was hiding.

Colonel Bob Jones was a "fair country lawyer" as one might expect of a good ol' boy parading as a "colonel." Warning: Watch out city shysters! Before the plaintiffs could get their two cents worth in, Jones requested the court to establish a voluntary receivership for Dad's leases so he could control the destiny of the ocean of oil he promised the folks of Rusk County. The country lawyer argued that Dad could not develop the well and leases nor protect the little investors with devious lawyers and Big Oil hanging over his shoulder. With flu-stricken and meek Dad sitting at the defendant's table, the judge agreed.

H.L. offered to buy Dad out before the pair left the courthouse, but Dad told him he'd only be buying a "pig in a poke." H.L.'s total cash amounted to $109. However, his friend, Pete Lake, an El Dorado haberdasher who had backed him on past ventures, agreed to take a 20% interest in the Joiner leases for $30,000 and advanced H.L. "walking around money." H.L. also obtained the services of H.L. Wilford, a lease hound and promoter friend of Dad's, by offering him $25,000 if he got Dad to the bargaining table and helped cement a deal.

With a little cash and the bad news that dry holes had been drilled to the northeast and southeast of the Daisy Bradford No. 3, H.L. was ready to negotiate. Hunt's well to the south was reported to have a "show of oil," nothing exciting. In truth, Jick Justice told H.L. he thought the well was capable of producing 100 barrels a day. As far as H.L. was concerned, the key was the Deep Rock Oil well to the west that Frank Foster had promised to keep Charlie Hardin advised of its progress. H.L. promised Foster $20,000 for telling him if the well was a producer before the news became public.

On the morning of November 25, 1930, H.L. and Dad embarked on what was to be thirty-six hours of almost non-stop negotiation in suite 1553 of the Baker Hotel in Dallas. No lawyers were present. Both oilmen knew attorneys have a way of screwing up a free lunch with paragraphs beginning "whereas," clauses starting "provided," and sentences sprinkled with Latin mumbo jumbo, such as *causa mortis* and *en ventre sa mere*. The session was a one-on-one, no holds barred battle between the Bible quoting wheeler-dealers. It mattered little neither had been involved in a million dollar deal before; they didn't have three months formal education between them; Dad didn't have good title to the leases; or H.L. didn't have the million dollars he knew was Dad's bottom price. Oilmen will find a way.

There are numerous conflicting accounts of what was said and occurred, but no one really knows. I cannot believe those who claim H.L. got Dad liquored up and entertained him with ladies of the evening.⑧ Dad did not drink the devil's brew. Moreover, Dea England was waiting for him across the street in the Adolphus Hotel, and the twenty-two year old was more than the seventy year old man could handle. What is known, is that during the negotiations, H.L. received information from Frank Foster on the Deep Rock well and three hours *before* the contract was signed, he was told the well had hit ten feet of the Woodbine sand saturated with oil.

Everyone seems to agree that the two men celebrated their deal with a plate of cheese and crackers, which should put to rest the ugly rumor that H.L. got Dad "liquored up." Dad's last words were: "Boy, I hope you make fifty million dollars."

Where did H.L. get the money to pay the purchase price of $1,335,000? H.L. only paid $30,000 down and gave promissory notes of $45,000 payable over nine months. The $1,260,000 balance was payable in *oil payments*. In case you don't know what an oil payment is, it is a promise to pay a sum *if and when oil is produced* from particular wells or leases. If Dad's leases did not produce the amount due based on 7/32 (first half) and 7/64 (second half) of the oil sold under the agreement, H.L.'s payment obligation ended. In other words, H.L. risked only Pete Lake's $30,000 investment and $45,000 in promissory notes, if you don't count Wilford's $25,000 "commission" and Fosters's $20,000 "consulting fee."

Two weeks later, Deep Rock's well gushed in at 3,000 barrels a day and H.L. announced his well in the south was producing 100 barrels a day. A major oil company offered Dad $3 million for his leases, but they were too late. H.L.'s Panola Pipeline,[16] named after the adjacent county to the east was being hooked up to the Daisy Bradford No. 3 and the Deep Rock well.

What did H.L. buy with Pete Lake's $30,000? According to his attorney, J.B. McEntire, over 300 lawsuits. Dad did not have clear title to two acres out of the 5,000 acres in his block of leases. Many lease titles were clouded with multiple assignments or burdened with liens and mortgages because Dad didn't have the funds to retain a lawyer to check the titles. However, titles and legal actions were a lawyer's problem. H.L. planned to drill and develop the leases and let McEntire worry about the mundane issue of whether he had a legal right to drill. H.L.'s instructions to McEntire were simple: settle the claim for a small amount or drag them through the court for years. Once he started paying the landowners royalties, they would settle.

H.L. was broke and needed $5,000 to cover immediate expenses until he lined up a bank willing to finance his venture. When he approached R.A. Motley, the banker advised that the Overton bank was overextended, but H.L. lucked out. Leota Tucker was in the bank to deposit a $30,000 check received as a bonus for leasing the family farm ten miles north of Overton. Land she couldn't sell at $30 an acre a few years earlier had reaped a $100 per acre lease bonus for the right to drill. Leota loaned H.L. $5,000 at eight percent. Shrewd Leota knew the interest rates ran less than three percent and insisted that H.L. continue to pay the full interest, even though two weeks later he tried to pay off the debt.

---

16 H.L. believed the Panola Pipeline Co. was a good omen and later gave all Hunt Oil affiliates six letter names beginning with *P*: Parade, Placid and Penrose.

On the strength of his holdings, H.L. borrowed $50,000 from a Dallas bank and arranged a line of credit. The Hunt Oil Co. was now in business. One of his first costs was to pay Frank Foster a "little on his $20,000 consulting fee."[17] Years later, H.L. would testify that he didn't remember why he paid Foster the money.

---

17 A consulting fee of $20,000 was a handsome sum in 1930 when the author's father was earning a good wage...$25 a week.

12
THE BLACK GIANT

Malcolm Crim had more reason than ever to believe the Gypsy's prophecy that there was oil under the family farm. However, no one else did. The leasing boom had not reached his farm fourteen miles north of the Daisy Bradford No. 3 and one mile south of the Gregg County line. Around Overton and Henderson oil leases were selling for between $500 and $1,000 an acre, but Malcolm's offer to lease the family farm for $2 an acre had no takers.

Following up on gossip that Malcolm was desperate to lease his farm, Ed Bateman, a poor boy wildcatter with a gift for blarney and promotion in the mold of Dad Joiner, went to Kilgore and made Malcolm an offer. Bateman had no money. He couldn't even pay the $2 an acre bonus Malcolm was asking for a lease. Malcolm reluctantly agreed to permit Bateman to drill without paying a bonus and assembled leases covering 1,500 acres of his and neighboring land.

Malcolm cringed when he saw it. Bateman's rig, designed to drill a maximum of 2,500 feet, was a duplicate of the junk Dad had used to drill the Daisy Bradford No. 3. Only able to raise a few thousand dollars for the risky venture, he traded small interests in the well for pipe. The first roughnecks hired glanced at the antiquated rig and left laughing. Undaunted, Bateman finally lined up a crew that would take a part of their wages for small fractions of the lease.

After drilling almost 3,600 feet, Bateman and Malcolm became concerned that there was no sign of the Woodbine. The Daisy Bradford No. 3 had hit the Woodbine at 3,486 feet. Even the best poor boy driller would find it impossible to drill much deeper with the third-rate equipment. The old-fashioned fishtail bit (called so because it looks like the tail of a fish), was only cutting an inch or two an hour through the hard formation.

A poor boy possibly close to a paying formation and without funds to rent a roller bit was a good time for an oil scout, like Charlie Hardin, to prove helpful. H.L. loaned Bateman a roller bit and core barrel on the condition he inform him if he struck oil before the news was public, allowing H.L. time to pick up cheap leases in the area.

The following day the roller bit ripped through 100 feet of hard cap rock and brought to the surface Woodbine sand cuttings dripping with oil. A week later Bateman brought in a 22,000 barrel a day gusher.

TEXAS BILLIONAIRE TRIVIA

Howard Hughes, Sr., the inventor of the roller bit still in use today, former poor boy wildcatter and partner in the Moonshine Oil Co., passed on to his son, Howard Hughes, Jr., a billion dollar company, allowing Junior to become a renown pilot, movie producer, womanizer and weirdo recluse.

As the oil rained down, Malcolm asked the driller: If a well could gush 20,000 barrels a day, could five wells produce 100,000 barrels a day? Advised it was possible, Malcolm ran into town to tell his mother that the well named after her, the Lou Della Crim No. 1, came in a gusher and they were rich.

Within days, leases between Daisy's and the Crim farm were selling for up to $5,000 an acre. But, It didn't end there. The Gregg County and Longview Chambers of Commerce posted a award of $10,000 to the person drilling the first producing oil well in the county. Three weeks later on the Gregg County farm of F.K. Lathrop, twelve miles north of the Lou Della Crim No. 1, a third well blew in at 20,000 barrels a day. The ecstatic driller and part owner, John Farrell, gave the $10,000 prize to the drilling crew and the biggest oil boom in American history took off.

The Black Giant

The Black Giant proved to be forty-five miles long and between five to thirteen miles wide. The East Texas oil field would eventual disclose it stretched from Rusk and Gregg Counties into Upshur and Cherokee Counties on the north and Smith County on the west. Lease prices soared to $15,000 an acre. Kilgore, in the center of the field, exploded from a town of 700 peaceful citizens to 10,000 boomers within two weeks.

Everyone from all walks of society rushed to take part in the oil business — farmers, politicians, teachers, store owners. The average well only cost $25,000 to drill and complete and an experienced driller with a modern rig could complete a 3,500 foot well in three weeks.[18] Due to the tremendous underground pressure, the well flowed freely without the need of pumps, saving the operator production costs. Within nine months of Daisy's discovery well, there were 625 oil operators in the field and the Black Giant was producing one million barrels a day. Cotton at 9¢ a pound held little interest to farmers. Oil rigs, equipment and supply roads soon took the place of cotton, corn and sweet potatoes. Landowners promoted their land in small parcels at oil exchanges set up in Henderson and Kilgore.

ASSASSIN TRIVIA

In the days of the Oil City, Pennsylvania, oil boom, actor John Wilkes Boothe formed the Dramatic Oil Co., but President Lincoln's assassin never struck oil.

Large tracts of land became almost impossible for the hordes of landmen to find. But, as far as little independent oilmen were concerned, large blocks of leases were no longer necessary. Poking a hole 3,500 feet in the ground was "like shooting ducks in a barrel," a wild-

---

18 One has to wonder what Dad Joiner did with the money he raised over three years and could have accomplished, if he provided decent equipment and pipe to his drillers.

catter declared. The misguided advice given to Malcolm Crim that five wells could produce five times a much oil as one well sounded mathematically correct. Malcolm and other store owners along the main street of Kilgore tore down the rear of their stores and set up derricks. One well was drilled through the marble floor of a bank and a church was torn down to make way for a derrick. Over forty wells sprouted on one block — billed the richest block in America. In many cases the derrick legs overlapped on the small lots until every landowner had his own oil well.

◆ ◆ ◆

Long before the discovery of the Black Giant, a basic law of oil and gas had been established — *The Rule of Capture*.[19] Simply put, oil and gas are fugacious and migrate across property lines, and whoever captures it owns it. Although the legal roots went back to the early oil industry days in Pennsylvania, the law was understood in Texas. It followed the oil patch version of the Golden Rule: *Do unto others before they do it unto you.* Everyone was hellbent to drill a well and get the oil out of the ground before his neighbors beat him to it.

Within a week of Bateman's Lou Della Crim No. 1 gusher, he sold his lease to Humble Oil for $2.1 million. John Farrell, part owner of the Lathrop well, sold his interest to a large independent for $3.2 million. Thus, the wildcatters, Dad, Bateman and Farrell were off wildcatting elsewhere.

Never content to settle down in one place, a wildcatter is a special breed of man, dreaming of the next gusher and exploring the unknown. His forbearers were the frontiersmen who explored America from the Alleghenies over the Rockies to the Pacific. Once they tamed the

---

[19] For the few lawyers who enjoy researching dry legal precedent and the many who must build up hourly fees to charge their clients, *see Barnard v. Monongahela Natural Gas Co.*, 216 Pa. 362, 65 A.801 (1907). The Pennsylvania Supreme Court held: "[E]very landowner or his lessee may locate his well wherever he pleases. *** What then can his neighbor do? Nothing; only go and do likewise. He must protect his own oil and gas. He knows it is wild and will run away if it finds and opening and it is his business to keep it home."

The Black Giant

wilderness, they moved on seeking new adventures. They forever seek the "big score," as noted in the apocryphal tale told by Samuel W. Tait, Jr., in *The Wildcatters*. A wildcatter who struck oil in Southern Illinois and refused to enter into a partnership with local merchants and a banker explained:

> Nothing doing! I know what I would be up against with those fellows...If we got a lot of little stinkers pumping a few barrels a day, I'd be expected to stay around and look after them as long as they paid a measly five or six percent. And all the while fields would be blowing in all over the Southwest and I wouldn't be there. No thanks!

First, Dad celebrated with Dea, then he started gathering leases for a wildcat play in West Texas, then he celebrated with Dea again. The unlikely pair were seen living it up in Dallas' best clubs and restaurants every night. Not only was Dad receiving oil payments every month, he had escaped going to jail for fraud, although few believe you could have found twelve jurors in East Texas who would have convicted the old con man.

The $30,000 H.L. paid Dad was more cash than the old man had ever seen that he could call his own and was spent faster than Dea could say "Neiman Marcus." It came as no surprise to H.L. when Dad showed up accompanied by Dea on many occasions, no one knows how often, for advances on oil payments of 7/32 of the oil sold.

An a typical visit, Dad would remark, "Boy, I'm on to something big in West Texas. Can you advance me twenty thousand on my next oil payment?" H.L. reminded Dad that he was breaking the rules — wildcatters don't risk their own money — and he should dust off his old sucker lists. Unfortunately, most of Dad's investors were suing the con man. While Dad cooperated in testifying in the cases, H.L. informed him that he wasn't responsible for defending the suit of Stella Sands, who claimed she was Dad's partner and entitled to one-half of Dad's interests in the field.

H.L. advised Dad that the oil payments would be less the next month. The price of oil had dropped and under their agreement Dad would only be entitled to oil payments of 7/64 in the future, half of what he had been receiving.[20]

A few months earlier, the two wheeler-dealers had been scrounging to make ends meet. Now they were writing checks for tens of thousands of dollars. However, as the price of oil dropped, their concern about the future became grave.

---

[20] Oil interest fractions often appear minute. The author once sold a pump jack worth $2,500 for a 5/512 interest in a lease. Two weeks later the poor boy brought the well in and sold the lease. My share was $8,000. (It could have been better. My interest would have earned an average of $6,500 a year during the next ten years.)

PART THREE

BIG OIL

The meek may inherit the earth, but not the mineral rights.
J. Paul Getty

13
BIG OIL V. LITTLE OIL

"**The world in running out of oil,**" the so-called experts claimed in the early 1920's. During World War I, the Bureau of Mines bureaucrats in the Department of the Interior had predicted that America would run out of oil in precisely nine years and three months. Oil not only turned the wheels of cars and industry, it was a matter of national security. British War Minister Lord Curzon's observation was quoted in the halls of Congress: "The Allies floated to victory on a sea of oil." Congress knew America produced 80% of the oil for the Allies victory over Germany.

The specter of a shortage, however, was soon forgotten. Although not as prolific as the Black Giant, enormous oil fields were discovered in Texas and Oklahoma during the Roaring Twenties. The world was being flooded with oil from Russia and the new oil fields in Persia and Mesopotamia (Iran and Iraq to those who forgot ninth grade history). The oversupply gave birth to fierce competition and its offspring, lower prices...the reason Big Oil met in Scotland to fix prices and eliminate competition.

Although America had an oil surplus, the multinational oil companies imported crude oil from Mexico and Venezuela. By 1928 Venezuela's vast Maracaibo oil field made it second only to the United States when it surpassed Russia. After the Barroso well blew in at

100,000 barrels a day, Gulf, Amoco and Exxon joined Shell, the discoverer of the Maracaibo field.[21] Independent oilmen also rushed to Venezuela to reap the huge profits, including William F. Buckley.[22]

Humble Oil, Standard Oil of New Jersey's (Exxon) subsidiary and the largest purchaser of crude oil in Texas, cut its price from $1.30 to $1.15 a barrel in January 1931, lowering the basis of world prices under the Achnacarry agreement. Humble seemed like the logical company to blame, it was the subsidiary of the largest remnant of Rockefeller's Standard Oil Trust and the leader of Big Oil.

STANDARD OIL BASHING

> Standard Oil had long been fair game, even on Broadway. Eugene O'Neill's memorable line from *A Moon for the Misbegotten* summed it up: "Down with all tyrants! God damn Standard Oil!"

It soon became "us agin' them" — independent oilmen against Big Oil. The large integrated oil companies controlled the pipelines, refineries, tankers, marketing outlets and overseas markets and oil production and had been in cahoots to control the world oil market since the 1928 Achnacarry Agreement. Through March and May 1931, the majors, who purchased 90% of the crude oil, dropped the prices they

---

21 In addition to Venezuela's enormous cheap flush production, the cost was reduced under government royalties as low as 7%, compared to the basic royalty of 12½% (1/8) in the United States that can reach 3/16 or 1/4 during an oil boom. Shell and Amoco, concerned the Venezuelan people might wake up some day and realize they were being screwed by Big Oil and their corrupt dictator, built their oil refineries in the Netherlands Antilles a few miles off the coast of Venezuela where they couldn't be expropriated.

22 Buckley's son, William F. Buckley, Jr., became a conservative publisher, TV commentator and needle in the backside of liberals and big government.

purchased oil from the independents to 67¢ a barrel, then to 35¢, then to 15¢. Desperate producers sold at any price in order to continue producing and empty their storage tanks, which led to sales of 10¢ and a few 2¢ desperation sales.

Big Oil continued to maintain their overall profits by keeping their refining, pipeline and marketing margins high. Drivers pulling up their cars to pumps in New York, Boston and Peoria reaped little benefit from the cutthroat oil prices.

The independents, not owning refineries, pipelines and gasoline stations fought back by drilling more wells. By June 1931 over 1,000 well dotted the East texas field and, by the end of 1931, the landscape was bristling with 3,600 wells. The Black Giant was producing one million barrels a day — one-half of America's demand for oil.

The Humble octopus couldn't blame it all on the small independents, it drilled wells faster than any other operator and strengthened its tentacles on the field by completing its pipeline to the field in February.

Congress was well aware of Rockefeller's methods of dominating the oil industry at the turn of the century through his Standard Oil Trust. Rockefeller learned he couldn't control the hundreds of independent producers, so he set out to control the transportation of oil. He made secret deals with the railroads to obtain cheaper rates and rebates on the tariffs and acquired the ownership of over one-half of the tank cars in the United States and the principal tank car manufacturer. When pipelines began to compete with rail transportation he gobbled up the key oil pipelines. After the Trust's transportation bottleneck squeezed the refiners with high transportation costs, it was able to grab control over 85% of the refinery capacity in the nation, giving it the wherewithal to dictate crude oil prices to the producers and fix the price of refined products to the consumers. Not only did the Trust refine gasoline for automobiles and kerosene for lamps and stoves, it manufactured paraffin candles and pharmaceuticals, the most famous being Vaseline.[23]

---

23 There is a joke in connection with Vaseline and the Standard Oil Trust's shafting the oil producers and consumers, but it was considered too risque for the genteel readers of this novel history.

In 1906 the Congress passed the Hepburn Act, which required all interstate pipelines to operate as common carriers and transport and purchase oil of competitors without discrimination. In holding the Hepburn Act constitutional, the Supreme Court singled out Standard Oil's duplicity and duress; however, the Act did not apply to intrastate pipelines. In 1930 the Texas legislature, upon learning that Humble was earning in excess of $1 million a month on its Texas pipelines and squeezing the independents' profits, passed Common Purchaser Act, requiring pipelines to take oil ratably and without discrimination from all producers. Humble, the largest pipeline owner in the United States, reacted by telling the legislature and independents to stick the law in their boreholes. (It's not what you think. It's a term for an oil and gas well drill hole.) Humble announced it was discontinuing the purchase of crude oil in several counties.

Many suspected the drop in price below 10¢ a barrel was not entirely due to the law of supply and demand, but Big Oil pushing prices down to destroy the independent oilmen who had arrived in the East Texas field before Big Oil.

What was the federal government doing about the mess in East Texas? "Nothing," according to H.L. Hunt.

During the early 1920's, the farsighted president of Cities Service Co., Henry L. Doherty,[24] preached that oil fields should be unitized and production controlled in order to conserve the gas and water pressure lifting the oil to the surface. In the beginning, he called his novel concept "rationalization" — it was rational that the nation not waste the precious resource. Unitization is simply all operators cooperating in the development and production of a common pool of oil in order to obtain the maximum amount of oil at the lowest cost. Doherty correctly argued that the drilling of unnecessary wells and uncontrolled production under the Rule of Capture left oil in the ground that might have otherwise been produced. Doherty was a *conservationist*. In addition, under field wide unitization, royalties would be fairly apportioned among the landowners.

---

24 Doherty was a visionary, but he never would have imagined that his progeny, Cities Service, which markets today as Citgo, would now be owned by Petroleos de Venezuela SA, the Venezuelan national oil company.

Doherty's pleas were ridiculed by Big Oil, including William Farish, Humble's president. Although a director of Big Oil's lobbying arm in Washington, the American Petroleum Institute (API), Doherty was refused permission to address its annual meetings on unitization in 1923, 1924 and 1925.

However, Doherty caught the ear of President Calvin Coolidge when he told him that twice as much oil could be produced if oil fields were unitized. Coolidge was leery about anything connected with oil. His Republican party was still reeling from the Teapot Dome scandal.[25] Silent Cal, never the one to waste anything, appointed the Federal Oil Conservation Board, composed of the Secretaries of Interior, War, Commerce and Navy, to investigate wasteful practices and its threat to national security. The Board did little except plead for conservation under Coolidge and even less under Herbert Hoover when new oil fields started blowing in, including Oklahoma's Greater Seminole field, which Dad claimed he would have discovered if he had not run out of money and had been able to drill another 150 feet. The Greater Seminole field produced 500,000 barrels a day by 1927. America had an oil glut!

As prices started to drop in 1928, Big Oil jumped on Doherty's conservation bandwagon. Even Humble's president, William Farish, saw the light. It wasn't a case of the Sierra Club or Friends of the Earth altruism to protect our nation's resources. Big Oil was out to cut oil production and raise prices. The API, Big Oil's Washington megaphone, which had argued that waste from not unitizing was negligible two years earlier, did an about face and endorsed Doherty's ideas. Unitization was supported by the American Bar Association, controlled

---

25 President Harding's Secretary of the Interior, Albert B. Fall, had illegally leased the Navy's petroleum reserves in Wyoming and California, in part, for $100,000 in cash delivered in what was to become synonymous with political bribes — "a little black bag." Fall went down in history as the first Cabinet officer convicted and sent to jail. Edward L. Doheny of Pan-American Petroleum, who bribed Fall, was acquitted, proving O.J. Simpson's theory that if you are guilty, hire the best lawyers. Unlike O.J., who didn't pay all his attorney's fees, Doheney paid a legal fee of $1 million...unheard of in 1930.

by fat cat business lawyers, including Charles Evans Hughes, a former Supreme Court Justice and later Chief Justice, who coincidently happened to be counsel for the API. The API also recommended United States crude oil production be held to 1928 levels, not coincidently identical to the secret Achnacarry "As Is" Agreement adopted in Scotland by Big Oil in 1928.

Independent oilmen smelled a rat. The endorsement of unitization by Humble's William Farish was its kiss of death in Texas. Fearful of government regulation and the API's power in Washington, they formed the Independent Petroleum Association of America (IPAA). One of the first impertinent questions the IPAA asked was: If there is an oversupply of crude oil in the United States, why was Big Oil importing crude oil from Mexico and Venezuela?

BIG OIL'S MOUTHPIECE

The API was a reorganization of the National Petroleum War Service Committee (Big Oil and a few large independent oil executives) assisting the U.S. Fuel Administration to supply oil during World War I. The NPWSC's chairman was...guess who?...the president of Standard Oil of New Jersey. After the war the NPWSC remained unified to wield power in Washington by forming the API with Mark Requa, the former head of the Oil Division of the U.S. Fuel Administration, as director. The API's first task was to contest the Federal Oil Conservation Board's unitization and conservation efforts.

Federal officials all the way up to the newly-elected savior of our country, Franklin Delano Roosevelt, claimed that they didn't Know Big Oil was engaged in any hanky-panky. If FDR had followed the wisdom of Will Rogers, "All I know is just what I read in the papers," he would have known that the independents and public were being screwed. On May 13, 1933, the *Petroleum Times* reported:

> *Disturbing news from East Texas where the increase in production alone in the past few days is many times greater than the total output in Rumania, and this formidable increase in production has been met by the international group posting crude prices down to ten cents per barrel <u>in order to kill the efforts of the independents</u>. (Emphasis added.)* [26]

The sweet oil boom turned sour. Independent oilmen went bankrupt when forced to sell their crude oil below the cost of production. Drastic cuts in oil prices from $1.30 a barrel to 10¢ meant a 92% drop in state severance taxes. The University of Texas Board of Regents became alarmed that the loss in royalties would cripple the funding of the state's educational system. The governor and legislature did not know how they were going to fund government.[27]

Texan's populist traditions against big business and the monopolies held by the railroads and oil companies was part of their undoing, but not without reason. After the War of Northern Aggression, Texans fought to recapture their banks and railroads from the damnyankee (one word) carpetbaggers and Eastern bankers. Until the Spindletop oil field discovery in 1901, Texas imported most of its oil from the Midwest and Appalachia, through the Waters-Pierce Oil Co., which controlled 90% of the market through underhanded practices and political bribes. When it was revealed that Waters-Pierce was 60% owned by Standard Oil, Texas revoked its corporate charter. Strange as it may sound today, *Texas was against anything big.*

---

[26] For those wondering what Rumania had to do with Texas oil...Rumania was ticked-off. It had signed the Achnacarry Agreement and was threatening to pull out. Russia, another signatory, had no gripe...the Russians always cheated.

[27] In 1986 when oil prices fell from $38 to $10 a barrel, it was estimated that for every $1 price decline, the State of Texas lost $100 million in taxes annually. It is little wonder Texas' congressional members, regardless of political party or philosophical bent, love the sweet smell of oil.

In 1919 the legislature enacted Texas' first petroleum conservation laws to prevent the physical waste of oil and charged the Texas Railroad Commission (TRC) with enforcement. Among the myriad of regulations was a provision to protect the "correlative rights" of the landowners by lessening the impact of the Rule of Capture. The TRC's infamous Rule 37 provided that wells must be located a minimum of 150 feet from property lines and 330 feet from the nearest well producing from the same formation. At first blush, the spacing rules would have made drilling on the lots in Kilgore illegal, as most were less than 100 feet wide. However, the rule also provided for exceptions for the owners of small lots under the populist theory that every landowner must have the opportunity to exploit the oil under his land. Loosely translated: In Texas, every SOT or DOT (Son or Daughter of Texas) was entitled to have one oil well, even if his plot of land wasn't large enough to hold a derrick.

The large landowners, principally farmers, became irate when the TRC attempted to limit the number of wells to one per ten acres, the area geologists believed one well could drain without causing "physical waste" — using too many rigs and the construction of unnecessary surface facilities, such as pipelines and roads. The oversimplified theory that limiting the number of wells would reduce the amount of oil produced also permitted landowners having small plots of a quarter of an acre to unfairly drain oil from neighbors owning parcels of ten or more acres.

"Umpires" were hired by the TRC's Central Prorationing Committee to determine the number of wells allowed according to the acreage in a lease, the well's production potential and whether to grant exceptions to the distance from the boundary and nearest well under Rule 37. As the TRC became overwhelmed with applications, it hired outside umpires, who were often employees of the major oil companies, which didn't come as a big surprise to the independents. Of the 24,000 wells eventually drilled in the field during the 1930's, 16,000 — two-thirds — were issued permits as exceptions to the spacing rules.

In most oil producing states, spacing rules based on the number of acres one well could effectively drain were adopted and pooling was required. Simply put, if a state conservation commission determined one well could efficiently drain the oil or gas from ten acres, all the landowners and leaseholders within the ten acres were required to "pool" their interests and share in the oil and royalties on a pro rata basis. Although Texas allowed voluntary pooling, the landowner's and oilman's fear of government and Big Oil resulted in Texas refusing to adopt compulsory pooling until 1965.

Texas independents and the landowners shuddered at Doheney's concept of unitization of an entire oil field now supported by Humble and Big Oil, except for the mavericks of the time, Gulf and Sinclair, who distrusted any form of government control. Today, Texas is the only oil producing state not enforcing compulsory unitization, although voluntary unitization is generally resorted to by the operators of an oil field for the purpose of secondary recovery after the normal methods of production are no longer feasible.

By now, you should be asking: What the hell does the Texas Railroad Commission have to do with oil?

14
GOVERNMENT — TEXAS STYLE

The Texas Railroad Commission (TRC) was the brainchild of Governor James S. "Jim" Hogg in the late 1800's. First as attorney general then as a two term governor, 300 pound Hogg pushed through and enforced antitrust legislation breaking the stranglehold damnyankee owned railroads had over agricultural and commercial freight rates and to make it illegal for out-of-state corporations to operate in Texas. Created in 1891, the TRC's authority was later expanded to administer corporate charters and enforce state anti-monopoly laws. In 1917 oil pipelines were added to its jurisdiction, as most were monopolies in individual oil fields. Thus, it seemed appropriate the conservation laws passed in 1919 to protect against the waste of oil and gas be handled by the TRC.

TRC & OPEC?

In an unofficial expansion of the TRC's duties and responsibilities, TRC Commissioner Kent Nance attended a meeting of the Organization of Petroleum Exporting Countries (OPEC) in 1988 to chat with the Arabs about how the TRC and OPEC could work together to prop up the low oil prices.

Hogg's name and booming voice were tremendous assets for an East Texas politician. Signs picturing a huge hog, slogans like *Don't Loosen the Belly Band*, and hog calls from supporters followed his campaign trail. When nominated for attorney general at the Democratic convention, he answered the delegate's hog calls to accept the nomination: "I am one of those unfortunate animals from the pine and persimmon valleys of East Texas that is not altogether a razorback, but I am glad to respond to your call."

The TRC was considered so important that Hogg convinced Senator John H. Reagan, to resign his seat in the U.S. Senate to be the first chairman of the TRC. Reagan, also from East Texas, had suffered imprisonment as a result of having served in Jefferson Davis' cabinet during the War of Northern Aggression and was intent on protecting the farmers and ranchers from the ordeal of Reconstruction repression, particularly railroad freight rates and monopoly controlled oil prices. As a fervent protector of the "little farmers of the west," he had introduced a bill in Congress that bridled the railroads — the Interstate Commerce Act of 1887. Reagan's first task in Texas was to rein in the intrastate railroads. The state's antitrust laws made it impossible for Standard Oil to operate in Texas until 1919 when it purchased 50% of the stock of Humble Oil. In 1909 Standard Oil's 60% owned subsidiary, the Waters-Pierce Oil Co., was fined $1,623,000 for operating in Texas and its charter revoked. Earlier, Texas was unsuccessful in its attempt to extradite John D. Rockefeller for violating Texas antitrust laws, but succeeded in convicting several Waters-Pierce officials of conspiring to restrain trade.

After the leaving the governorship, Hogg became known as a tough and well-connected lawyer and businessman, and bragged proudly: "Hogg's my name and hog's my nature." He despised Standard Oil and was a fierce opponent of Prohibition, which may explain why he named his daughter "Ima."

When the huge Spindletop oil field outside Beaumont was discovered in 1901, some called it "Swindletop," as more money was made selling leases and stock than oil during its early years. Hogg and his cronies obtained a 10% percent interest in the 240 acre field through a prudent investment or phony deal, depending on who you believe.

Gulf Oil, controlled by the Mellons of Pittsburgh, had bought out the interests of the wildcatters who discovered the field, but Gulf was still small and did not desire to battle Standard Oil. It was later explained by a Gulf Oil official, "Northern men were not well respected in Texas in those days, and Governor Hogg was a power down there." In other words: Gulf was a nice Pennsylvania oil company unlike nasty Standard Oil from New Jersey...keep Standard Oil out of Spindletop.

Hogg's "$105,000 investment" became known as the Hogg-Swayne Syndicate, as Hogg and his principal partner and former floor leader in the Texas Senate, Jim Swayne, were opposed to corporations. Immediately, they sold 2 1/2 acres of the 24 acre lease for $200,000. Then they realized they had sold *too low*! The next sale of 1/24 of an acre sold for $50,000. Soon they were selling "doormats" of 1/16 and 1/32 of an acre, the cheapest to a judge friend for $15,000, until 12 acres were studded with 120 wells. Crowding ten wells per acre led to the legs of the derricks overlapping, disastrous fires and the tragic wanton waste of the gas pressure, which curtailed oil production long before the original Spindletop oil field should have ceased harvesting the resource.

The cash rich Hogg-Swayne Syndicate decided to enter the oil business with the remaining 12 acres in the lease, but faced a big stumbling block — they didn't know anything about the oil business. They ended up swapping a few acres in exchange for $25,000 in stock in the Texas Fuel Co., run by "Buckskin Joe" Cullinan, a transplanted Pennsylvanian from Corsicana, Texas, and financed by a group from Chicago led by the infamous John W. "Bet a Million" Gates, that was to change its name to the Texas Company, then Texaco.

In deference to Hogg, Cullinan chartered the Texas Company in Texas, aware of his antagonism against foreign corporations. As Hogg, like most Texans, was opposed to integrated Big Oil and the Texas Company's original charter was limited to oil refining, Cullinan organized an affiliate, the Producer's Oil Co. for exploration and drilling. One has to wonder what Hogg, the populist who despised and fought big corporations, would have thought if he lived long enough to see Gulf and Texaco emerge from Spindletop to become members of the Seven Sisters and Big Oil.

PRODUCER'S OIL CO. LEGACY

Producer's Oil is said to be the first oil company to use a printed form oil and gas lease. Its simple one page eliminated haggling over terms, typing, review by lawyers and told Texas farmers that it was a "standard" oil and gas lease. Originally styled *Producers 8*, soon other oil companies followed with their version of a standard form called a *Producers 88*.

Today, over one-half the oil and gas leases are captioned *Producers 88*, but they are far from standard or identical as the Texas court pointed out when comparing the forms. *Fagg v. Texas Co.*, 57 S.W. 87 (Tex.Com. App. 1933.)

Ross Sterling, the newly elected governor, was a founder and former president of Humble Oil. He was also the owner of the Houston *Post-Dispatch*, which continued to say nice things about him when he screwed up and fronted for Big Oil. When he took the oath of office in January of 1930, Dad Joiner had yet to discover the Black Giant. However, it didn't take the independent oilmen long to figure where the governor stood when it was disclosed that Humble had paid him "advance royalties" of $250,000 and there was no fiscal appropriation for the enforcement of the Common Purchaser Act requiring Humble and other pipelines to operate as common carriers.

Humble pressed its former president to interpret the TRC's duty to prevent waste to include *physical and economic* waste under the state's conservation laws. Everyone understood the prevention of physical waste as a sound conservation practice — obtaining the greatest recovery of oil at the lowest cost of extraction. Economic waste was quite different — producing oil in excess of the market demand. The TRC's task was formidable. The legislature had amended the conservation laws in 1929 to provide that waste "shall not be construed to mean economic waste."

Ignoring the intent of the legislature, on August 14, 1930 — seven weeks before Dad Joiner struck oil — the TRC issued an order limiting the total oil production in Texas to 750,000 barrels a day based on its estimate of market demand. Most independents and many majors believed the order was illegal and paid it little or no attention and the TRC did not have the manpower to enforce its order, which called for "prorationing" — reducing the production of wells on a pro rata basis to meet the demand limitation.

When the East Texas field was capable of producing one million barrels a day — more than the total limit for the state — the U.S. District Court invalidated the TRC's order, ruling that the Commission exceeded its authority. The federal judge recognized the TRC's order was simply old-fashioned price fixing: "[U]nder the thinly veiled pretense of going about to prevent physical waste the Commission has, in co-operation with persons interested in raising and maintaining prices of oil...set on foot a plan which was seated in a desire to bring supply within the compass of demand."[28] Cautious independents who had lowered their production turned their wells wide open loose. Texas was long way from solving its problem of too much oil.

H.L. Hunt and many other independents realized there had to be a compromise or their wells and leases would become worthless and gobbled up by Big Oil. He had witnessed the rape of oil fields through overproduction in Arkansas and Louisiana in the ten years he had been in the oil business (since the Klu Klux Klan threatened to burn down his gambling hall in El Dorado).☺ If something was not done to curtail production, the East Texas field would die before its time because the needless waste of gas and water pressure would leave nature's gift unobtainable thousands of feet below the surface.

No longer a lease hound or out to promote a quick buck and run, H.L. was in East Texas to stay and ready to compete with big boys. His Panola pipeline to the railroad was running at capacity and netting a profit. It was time to become an integrated oil company like Big Oil. He purchased a small refinery and, in keeping with his superstition of six-letter company names beginning with *P*, named it Parade.

---

28 *MacMillan v. Texas Railroad Commission,* 51 F.2d 400, 405 (W,D.Tex. 1931), *rev'd per curium and dismissed,* 287 U.S. 576 (1932) (moot after the law was amended).

H.L wasn't the only one to think of building a refinery and manufacturing gasoline instead of selling his crude oil at 10¢ a barrel. The Black Giant was blessed with sweet light crude oil ranging between 37 to 41 API° gravity. A teakettle refinery costing $10,000 to $25,000 to construct could produce 16 to 18 gallons of gasoline from a barrel of crude oil. With refining and marketing costs of approximately 20¢ a barrel, the gasoline production alone (not counting fuel oil and small quantities of kerosine) could be sold at 4¢ to 5¢ a gallon to gasoline stations and raised the per barrel profits to between 44¢ and 70¢. A simple distillation teakettle perking 2,000 barrels a day would net between $880 and $1,400 a day and payoff the construction costs in one month.

"Eastex" gasoline could be sold for 10¢ or 11¢ a gallon at gasoline stations far below the major's pump price. Soon there were over seventy-five teakettles topping off Eastex gasoline and shipping it by tank truck across the state line. Small refiners competed against Gulf, Texaco, and Standard Oil's offspring outside of Texas, such as Standard Oil of Indiana (Amoco), as well as Humble. Although Eastex gas extracted from crude refineries carried no octane guarantees, at 10¢ a gallon it was a welcome bargain to the motorists during the Depression.

CONFUSED ABOUT GALLONS & BARRELS?

Although barrels have long since disappeared from the oil fields, oilmen still compute a barrel of oil as containing 42-gallons and will tell you the *Exxon Valdez* only spilled 260,000 barrels in Alaska's waters. Environmentalists will cry that 11,000,000 gallons fouled Prince William Sound's pristine waters because its sounds a lot worse.

In 1866 Pennsylvania oilmen adopted the 42-gallon standard English barrel fishermen had used to pack herring since King Edward IV established a law in 1482 to prevent hanky-panky and "divers deceits." The Americans considered the standard 36 gallon English beer barrel, but it was rejected as too small and it might create a shortage of beer or make their beer taste terrible.☺

Oil Spill Trivia
The Exxon Valdez was not the worst oil spill! That distinction goes to the *Amoco Cadiz*, which ran onto the rocks off the coast of France in 1978 and dumped approximately 1,605,000 barrels — a whopping 67,200,000 gallons according to the Sierra Club.

Tanker Measurement Trivia
As all crude oils have different weights, tankers, such as the *Amoco Cadiz,* weigh their cargo in long tons (2,240 pounds to distinguish it from the American 2,000 pound short ton). Thus, it should come to no surprise that Amoco said it only spilled 223,000 tons of crude oil from its 1,095 foot supertanker over three football fields long.

Petroleum Measurement Fact
Measurements of barrels of oil and gallons of gasoline at the pump are adjusted to a temperature of 60°F (15.6°C) to compensate for the expansion and contraction of petroleum with the temperature.

Oklahoma also felt the price squeeze. Governor William "Alfalfa Bill" Murray ordered the National Guard to close down the Greater Seminole and Oklahoma City fields. More honest than Governor Sterling, he vowed the fields would remain shut in until oil returned to a "dollar a barrel." On August 4, 1931, the four oil wells on the lawn of the governor's mansion stopped pumping.

Governor Ross Sterling vacillated, like most politicians handling an explosive issue that could swing the next election. First he took the pulse of his constituents, then:

> On August 12, 1931, at the Governor's urging, the Legislature passed the Anti-Market Demand Act prohibiting the TRC from prorating oil production on the basis of market demand.

On August 13 the news broke that dozens of East Texas operators in favor of market demand proration sent Alfalfa Bill Murray a congratulatory telegram on his action and commented they regretted their governor hadn't shown the same courage.

On August 14 the East Texas Chamber of Commerce sent Sterling a petition with over a 1,000 signatures requesting that martial law be declared in Rusk, Gregg, Upshur and Smith Counties.

On August 16 Governor Sterling issued a proclamation declaring the existence of "a state of insurrection against the conservation laws of the state...in open rebellion against the efforts of the constituted civil authorities in this state."

On August 17 Governor Sterling ordered the National Guard to close down the East texas oil field.

15
HOT OIL

Blustering clouds gathered overhead as the long train slowly grinded to a halt. The storm foretold the future of East Texas and portended the passions of its people for years to come. Minutes later, 1,200 Texas National Guard troops disembarked in Kilgore to enforce marshall law and shut in the Black Giant.

The military contingent came in force with infantry, cavalry and several World War I aircraft. "Hell, they even brought a band," one hooker yelled. Headquarters were set up in the Kilgore City Hall and the troops bivouacked near where the Lou Della Crim No. 1 was drilled. Conflicting theories arose why the troops were stationed on the lease now owned by Humble Oil, which would earn it the name "Proration Hill." Some joked it was because Humble was the biggest prorationing cheater. Others swore it was to protect Humble from the angry independents.

Politically and morally, Governor Sterling could not have made a worse choice of commanding officer. Brigadier General Jacob F. Wolters was a pompous ass who loved playing soldier and parading in his campaign hat and billowing riding pants. His cherub face and paunch would have looked comical were it not for beady eyes and a mean thin mouth held tight by the strap of his wide brimmed hat. Looking as mean as a snake did not phase East Texas boys and oil patch roughnecks, many could stare down a wildcat and bite its ear off. What "riled 'em up" was Wolters was the general counsel of Texaco — Big Oil that had left Texas for New York — and Wolter's adjutant was Walter Pyron, a Gulf Oil official.

To most small independent oilmen and farmers dependent on royalties, it was an army of occupation laughingly called "Boy Scouts." However, Wolter's reputation preceded him. He had led the troops during Longview's racial riots, broken violent dock strikes in Galveston and was credited with bringing "law and order" to the oil boom towns of Mexia and Borger. The fact that he was soundly defeated when he ran for the U.S. Senate indicates he wasn't loved by the majority of Texans.

Given a free hand by the Governor, Wolter banned meetings of over twenty-five citizens in Rusk, Gregg, Upshur and Smith Counties. The general also barred hookers wearing brightly colored beach pajamas on the streets while extolling their wares on the basis that the nearest beach was over 100 miles away. But he didn't ban prostitution. His troops were some of the "painted ladies" best customers and the women were expected to pay a "weekly $20 fine" to help defray the cost of Kilgore's new police department that had grown from one officer to over twenty during the boom.

After the general mounted the steps of the Kilgore City Hall and read the governor's proclamation declaring martial law because the area was in a "state of insurrection" and "open rebellion," the crowd laughed. Wolters left the steps to the blare of a military band that was to march through the towns for sixteen months playing band music for the entertainment of the citizens and as a reminder of military occupation. As you might expect, the band's favorite tune was *The Yellow Rose of Texas*.

YELLOW ROSE OF TEXAS

The Yellow Rose of Texas is said to have been a black slave, named Emily, and touted as having kept Santa Anna busy in his tent with sexual favors while Sam Houston snuck up on the Mexicans and whipped their butts. The line, *She's the sweetest little rose that Texas ever knew*, originally sang, *She's the sweetest rose of color, this darky ever knew*. (This is a good example why the myths and truth about the East Texas oil field, or anything else about Texas, is difficult to untangle.)

The boom became a bust. Thousands of boomers who fled the poverty of the farms and small towns and the bread lines of the big cities were forced out of work. Local landowners, whose royalties were tied to oil prices and had seen their incomes dwindle, now saw them dry up entirely when the oil field was shut in.

In the meantime, major oil companies and the large independents were buying up leases from the cash short small independents. Although martial law had shut down oil production, the proclamation was crafted to allow the drilling of new wells. Big Oil, with the financial ability to continue drilling, continued to sink hundreds of new wells.

There are generally two sides to an argument, but when it came to prorationing, there were dozens of perspectives and the independents couldn't agree on an answer. Humble Oil's William Farish, realizing his support for prorationing would do more harm than good, kept a low profile while directing the attack behind the scenes. The balding, bespectacled lawyer was as heartless as his Standard Oil predecessors and not trusted by the independents supporting prorationing. They knew Humble was keeping prices low to squeeze out the independents. No doubt he would have been driven out of Texas in a hearse if the independents were aware he had advised Walter Teagle, president of Standard Oil of New Jersey, that only the shock and pain of 10¢ a barrel prices would bring the independents in line for prorationing and unitization, and that Standard's only course was "the law of the tooth and claw." Teagle approved, it was tearing a hole in the giant's Texas pocketbook, even though Standard Oil was selling the crude oil purchased in the East Texas field at 10¢ or 20¢ a barrel for many times the price to Europe.

With the troops in place on September 5, 1931, the TRC issued a prorationing order limiting production in the East Texas field to 400,000 barrels a day and set an allowable of 225 barrels a day for each well. On October 28 the federal district court, sitting with a three judge panel because of the challenge to the constitutionality of the order, enjoined the TRC from enforcing the order. However, the joy in the oil patch was short. Governor Sterling and General Wolters brazenly refused to obey the injunction even when cited for contempt of court. The governor personally took over issuing prorationing orders cutting the allowable to 165 barrels a day, then 150, then 125, then 100, all in derogation of a federal court order.

On February 25, 1932, a prorationing order limited the field to 325,000 barrels a day and a 75 barrel a day per well allowable. Wells that once gushed 20,000 barrels a day were reduced to a trickle of their potential. In October 1932 the federal court ruled again that the TRC was actually basing its prorationing order on market demand and economic waste without authority of the legislature and that a per-well allowable was unreasonable in that it did not consider the acres drained or the well capability. Without the legal ability to enforce prorationing under the current conservation laws, the governor called the legislature into a special session.

After martial law enforced the allowable production, Big Oil allowed the price to creep up to 85¢ a barrel. The governor and prorationists pointed to a recent Supreme Court decision upholding Oklahoma's economic waste prorationing statute as a constitutional exercise of the state's police powers. Oklahoma, using finesse in crafting its statute, had stated that the purpose of the prorationing orders was to protect the "correlative rights" of the owners; that is, the opportunity to obtain the oil beneath their lands before it is drained by another. The Supremes believed the fairy tale that none of Oklahoma's orders "had been made for the purpose of fixing the price of crude oil or has had that effect."☺[29]

On November 12, 1932, the Texas legislature passed the Market Demand Prorationing Act. Farish's law of the tooth and claw had won the battle, but the war was far from over. The sentiments of the small independents who saw their dreams being sucked down the borehole to be gobbled by Big Oil was expressed by Texas Senator Joe Hill:

---

[29] As a law school professor who taught oil and gas law for fifteen years, the author is entitled to denote the language of the Supremes with a ☺. Governor Alfalfa Bill Murray had declared publicly during the district court hearing that his intention was to return to "dollar oil." Nerd lawyers should recall the Chief Justice was Charles Evans Hughes, the former API counsel. *Champlin Refining Co. v. Oklahoma Corp. Coms.*, 286 U.S. 210 (1932).

> It is the rankest hypocrisy for a man to stand on this floor and say that the purpose of prorationing is anything other than price fixing. I sit here in utter amazement and see men get up and talk about market demand as an abstract proposition, and contend that it has no relation to price fixing.

Governor Sterling was defeated for reelection by Miriam A. "Ma" Ferguson. Before leaving office, the Supremes slapped him in the face by ruling his proclamation declaring the East Texas oil field was in open rebellion and insurrection was a subterfuge and unconstitutional. The governor's proclamation that the oil field was in a "state of insurrection, tumult, riot, and a breach of the peace," was found to be a lie in the court's unanimous opinion written by Chief Justice Charles Evans Hughes.[30]

The independents watched oil prices fall as more wells were drilled and Big Oil retaliated against the independents. Most agreed that the oil field that they discovered had to be protected and nurtured as Henry L. Doherty advocated, but it had to be done fairly. They were not about to surrender control to Humble and its Big Oil sisters. However, they couldn't agree on enough issues to unite.

A frustrated TRC was unable to satisfy all the operators, as evidenced by the seesawing of the East Texas field allowable production from a low of 200,000 barrels a day up to 750,000. Shackled by politics and incompetence, the TRC was unable to placate anyone. And its corrupt umpires favoring the majors and bribed by the independents didn't help generate confidence among honest oilmen.

Populist politics also confounded enforcement. The Market Demand Prorationing Act required the TRC to "take into consideration and protect the rights and interests of the purchasing and consuming public of crude oil and its products." The general safeguard was in lieu of a myriad of amendments to protect public and special interests, such as the proposed amendment by senators from the cotton districts prohibiting the TRC from restricting oil production when the price of gasoline exceeded the price per pound of cotton. In 1931 the small

---

30 *Sterling v. Constantin*, 287 U.S. 378 (1932).

operators and land owners had convinced the populist legislature to pass the Marginal Well Act (MWA) exempting wells producing 40 barrels a day or less from well-spacing and prorationing. In theory, the MWA prevented waste by allowing production from wells that might be abandoned and leave the oil in the ground. With several thousand wells producing less than forty barrels, it made a mockery of the total allowables. In 1933 the legislature revised the statute to reduce so-called marginal well in the East Texas horizon to 20 barrels a day.

STRIPPERS

Today, marginal or "stripper" wells are those producing ten or less barrels per day. They make up roughly 70% of the wells in the United States, but only 15% of the oil produced. (In the Mid East, the average well gushes about 8,000 barrels a day.) Economists, conservationists and oilmen have differing views on the exceptions and benefits bestowed on stripper wells. Cold-hearted economists believe stripper wells create waste rather than prevent waste and legislation like the MWA fosters high-cost production at the expense of low-cost production and higher consumer prices.

My favorite stripper well fiasco occurred during the Department of Energy price controls of the 1970s after the Arab oil embargo when U.S. crude oil prices were capped a $7 a barrel, but permitted stripper well production to be sold at $14. A small Oklahoma operator, who had excelled in third grade math, owned 15 wells that produced and average of 11 barrels a day, grossing $1,155 per day. When he lowered his per well production to 9 barrels a day, they became stripper wells and grossed $1,890. So much for government controls.

Another problem facing the TRC was the statewide spacing rule of ten acres under Rule 37 adopted in 1919. The ten acres the TRC deemed adequate for one well to drain the oil was subject to exceptions, which constituted two-thirds of the wells drilled until the average well density in the East Texas field reached one well per five acres, twice the number of wells they thought necessary to obtain the maximum efficient rate.

Prohibition had its bootleg booze and General Wolters had bootleg oil produced in violation of the TRC's prorationing orders called "hot oil." The story goes it acquired its name one chilly rainy night when a National Guardsman mentioned he was cold to an oil bootlegger, who told him to lean against a storage tank recently filled with illegally produced crude oil from deep in the hot bowls of the earth and found he could warm himself with the "hot oil."☺

In the beginning, it was simple for the hot oilers to evade the National Guard cavalry riding through the woods in the mud of East Texas and its two antiquated planes flying overhead. In addition to taking a potshot or two at the planes, wells were hidden in barns, houses and thickets. General Wolters soon learned to also keep an eye on the small refiners who were buying hot oil. When the refineries were caught, the hot oilers shipped the oil out of Texas through the woods at night by truck where the Texas Rangers and National Guard had no jurisdiction.

The hot oilers were not only made up of desperate small independents fighting to survive, some were simply crooks out to make a fast buck who didn't pay the landowners their royalties. There were also thieves stealing oil by tapping pipelines. Nor were the small refiners producing Eastex gas the only ones purchasing hot oil sold below the posted price. A large percentage was purchased by major oil companies and large independents. Soldiers drawing $2 a day and railroad workers paid $4 for a ten hour shift were easy targets for bribes to hook up extra cars to a train pulling out of East Texas, most of which went to majors buying hot oil below the posted price. The majors and large independents helped perpetuate the hot oilers. Hot oil produced between 1932 and 1934 was far more than the Eastex teakettles were capable of refining...another way the majors kept the posted price low.

In 1932, when the Hunt Oil Co. had 145 producing wells in East Texas, the company was convicted and fined $49,000 for producing hot oil. H.L. denied he knew anything about it and blamed it on employees who were stealing from the company.☺

POSTED PRICE

Posted prices are prices purchasers publish they will pay for crude oil in an oil field. Refiners and pipelines, such as Humble, posted them publicly to advise the sellers of the current price, which also had the effect of advising other refiners or pipelines and invited price fixing. Of course, that didn't mean the refiner or pipeline didn't squeeze the little producer on occasion and pay less than the posted price.

For obvious reasons, no one knows precisely how much hot oil was produced. (Does a bank clerk admit how much he is embezzling?) The most quoted figures are those reported in the Bureau of Mines *Mineral Yearbooks*. However, it must be kept in mind that *important* statistics were gathered by bureaucrats who needed *important* facts to make them *important*, not to mention producing *important* analysis demanded by an egomaniacal *important* Secretary of the Interior.

EAST TEXAS HOT OIL
Production (millions of barrels)

Year	Total	Hot Oil	Percentage
1931	105.7	3.3	3.1
1932	120.4	25.3	21.0
1933	171.8	35.1	20.4
1934	158.4	22.9	14.5
1935	176.3	12.5	7.1
1936	160.4	2.8	1.8
1937	169.0	1.5	0.9

Hot Oil

Tom Patton went down in history as East Texas' most famous hot oiler. In London, Texas, a wide spot in the road later replaced by New London, Tom drilled three wells on the main street and connected them to valves in a house he built nearby where he could control the flow of oil unnoticed. He claimed the house as a homestead under Texas law, which required the National Guard and Texas rangers to obtain a search warrant each time they checked his well, earning him notoriety in Ripley's *Believe It Or Not.*

To transport the oil he built an underground pipeline to a refinery at night, which worked until the National Guardsmen became suspicious where the refinery was obtaining large volumes of oil. The sneaky soldiers found it with a metal detector and bulldozed the line. But Tom was sneakier. He installed a fire hose the soldiers were unable to locate with metal detectors until the soldiers, thinking Tom was smiling too much for their liking, dug a ditch around the house and found the hose. Tom's next move was to buy a store across the street and dig a deep tunnel from the store to the house and run a new fire hose to the store then to the refinery.

By the time the TRC discovered his last ploy, the Robin Hood of East Texas had run over one million barrels of hot oil. During the charade, the federal and state courts had issued three injunctions against the TRC, Governor Sterling and General Wolters declaring the prorationing orders illegal, all of which were ignored by the governor and general who thought they were above the law. The governor, TRC and Wolters called him an outlaw, but it was the governor, TRC and Wolters who broke the law, not Tom. When prorationing was upheld by the courts, Tom obeyed the law...or so he said he did.☺

16
OIL PATCH TRASH

Kilgore's population multiplied by fifteen within two weeks of the oil strike on the Crim farm. Families lived in unheated cars by the side of the road. When the towns couldn't hold the masses, they swarmed like locusts to cornfields. Tent and shack cities erupted like festering boils along the highways. Unlike the "Hoovervilles" of the displaced poor living in the squalor of cardboard shacks surrounding cities, oil patch shantytowns housed stores, hotels, restaurants, barber shops, saloons and whorehouses.

The crudely constructed towns didn't offer plumbing. Streams yielded only silted water for washing. Drinking water wasn't available unless an enterprising farmer with a mule hauled in a tank load to sell at 10¢ a gallon. Outside the established towns there was no electricity. Over a decade would pass before the Rural Electrification Administration would bring electricity and lighting to the country and farmlands. Kerosine stoves and lamps provided the amenities of heat and light, but were fraught with hazards in the rural ghettos. Joinerville, the first sprawl of shacks, burned to the ground. It's only trace is a sign on the highway extolling where it once stood.

To add to the misery, the constant deluge of rain after years of drought turned the unpaved streets and fields into quagmires and swamped the shacks. On the bright side, it fostered the return of occupations the automobile had rendered obsolete. Roads melting into strips of chocolate pudding bogged down the biggest and most powerful trucks. Trucks lugging heavy oil rigs and drill pipe sank to their axles in the corn and cotton fields. Mule skinners replaced truck drivers. Horses and oxen pulling wagons displaced trucks. Stables importing and selling mules and horses from as far off as New Mexico and Kentucky couldn't meet the demand. Old men who once plied the blacksmith trade came out of retirement and would earn more in the next few years than their previous work life sweating over a forge.

New and novel occupations flowered. Men and boys stood on Kilgore's wooden sidewalks waiting to carry ladies and gentlemen across the foot deep rutted mud for 10¢. Shops selling boots thrived as did shoeshine boys. Outhouse toilets in alleys were available for 10¢, children 5¢. On country roads, farmers laid planks and logs over muddy ruts and charged a $1 toll. More than a few farmers were accused of plowing up the road at night to keep it impassable. Towns and counties hired extra police whose sole job it was to levy fines on overweight trucks to fill the local treasuries.

EAST TEXAS MUD

You can't gripe about the mud if you prayed for rain. East Texas saying.

With any oil boom comes oil patch trash. Local citizens were appalled at the roughnecks, carpenters, bricklayers, machinists, cooks and the unskilled streaming in with their families seeking work and degrading their towns. It was one thing to have to cope with the local poor, now the destitute from Oklahoma, Kansas and even the damnyankee states were moving in and crowding the streets, schools and stores. But what frightened them most were the gamblers, prostitutes, pimps, con men and thieves who follow the boom town trail. Most towns couldn't afford but one or two policemen and were unable to control prostitution, theft, fraud and murder. Sin cities, such as Pistol Hill and Newton Flats, sprung up on Kilgore's outskirts to take the roughneck's and workingman's hard earned money as well as big scores from the high-roller promoter and wildcatter. Fundamentalist East Texans saw every vice engulfing their quiet communities — gambling, bootleg whiskey, prostitution and honky-tonk girls at a dime a dance. For $2 you could go to a back room and do more than dance. Stolen oil drilling equipment in short supply was sold openly by thieves to unscrupulous oilmen.

Lawless Pistol Hill was well-named. Thieves roamed freely and bodies stripped of money and jewelry were found in alleys in the early morning. By noon the victim's jewelry could be purchased in one of Pistol Hill's many hockshops. As one old-timer commented: "Pistol Hill weren't no place to take your sister."

Friday was payday in the oil fields. Roughnecks and roustabouts were more than likely to leave a good part of their wages in places like Pistol Hill. Many ran to the saloons and ordered a 5¢ beer in the grimy coveralls they had sweated in during their twelve hour shift. As they swigged the generally warm brew, they tossed dimes and quarters on a roulette table. Around them lay poker, faro, blackjack and crap tables they could belly up to and bet everything in their pocket. If they won a few dollars, they could count on a ruby-lipped hooker to mosey next to them and earn her share of his winnings.

The gambling halls didn't care whether you were a wildcatter, promoter or roustabout as long as you had money, unless you were Mexican, Chinese or "colored"...these folks had their own places where they could be skinned of their earnings. In the main room the fancy dressed and working man in overalls stood shoulder to shoulder betting against the odds and often a stacked deck or wheel. There was always a back room for the high-rollers where they could bet handfuls of $100 bills in a no-limit crap game.

There was no shortage of musicians from cities and country hamlets. Music came in two styles: Jazz blared from bands or pounded out by a ricky-tick piano player trying to be heard over the din of the revelers. More often fiddles screeched a lively country ditty or the twang of a guitar and a voice wailed over a girl who jilted a cowboy or a train chugging off to nowhere.

No one recognized the Mexican when he ambled into the saloon one rainy night. A plump, over the hill hooker winced at the smell and decided she didn't want his business. He hadn't shaved in days and his shirt reeked of sweat. Not even the biggest, toughest or drunkest roughneck looked him in the eye. A Colt forty-five sticking from his belt said he meant business and his black eyers burned as he studied the faces around the bar, obviously looking for trouble or someone. The bartender was relieved after he sipped half a warm stale beer, slapped the chubby hooker on the butt and sauntered out the door.

The Mexican's next stop was a fancy honky-tonk. He could tell it was high class, the blonde was singing in off-key New Orleans French and the bartender gave him the *we don't serve Mexicans* stare. After bulling his way to the bar through fancy dans and expensive ($10) hookers, he order a beer. When the bartender paid him no attention, he laid his Colt on the bar to improve the bartender's hearing. The Mexican laid a dime on the bar for the quarter beer (it was cold) then left. ☺

☺ ☺ ☺ ☺ ☺

The Crims were respected as one of the most honest and generous families in Kilgore. After striking it rich on the family farm and their downtown store lots, they cancelled the debts owed by their customers in celebration of their newfound wealth. As Malcolm Crim stood outside the door of the Crim's general store, shortened to allow an oil derrick in the rear, he admired the growing town, but cringed at its seamy side. Painted women strolling by and winking and sharpies diddling people with the age-old shell game on the street shouldn't be permitted in a decent town. As the newly elected mayor, Malcolm was going to do something about it he told the town officials standing next to him.

As if on cue, a tall swarthy rider on a black horse appeared on the street. Sitting erect, his white Stetson set squarely on his black curly head. Two pearl-handled Colts on his hip and a Winchester rifle jutting from a saddle holster brought a smile to Malcolm. The Mexican, now clean shaven and wearing a starched Western shirt, guided his horse through the crowd of awed onlookers.

"Gentlemen," Malcolm said to the men, "I'd like to introduce Sergeant Manuel T. Gonzaullas of the Texas Rangers."

"She-it," cried a East Texan, "Lone Wolf Gonzaullas!"

At thirty-nine, Lone Wolf Gonzaullas was already a legend. Fast with his Colts, Texas newspapers had chronicled his exploits from one boom town to another cleaning up lawless border towns, enforcing Texas justice with his guns and fists. Those foolish enough to resist ended up in jail or a pine box. He gave credence to the Texas cliché, *One riot, one Ranger.*

Lone Wolf's personal sense of justice did not include arresting brawling roughnecks, honest gamblers, bootleggers or ladies of the evening, the latter he described as merely providing "commercial romance." Any ordinary cop could handle petty crime. He would have enough on his hands ridding the area of gunmen, thieves and swindlers. The American Civil Liberties Union would not have approved of his practices. Posing as a mean, gun-toting Mexican for a week, he had wandered the saloons and worst parts of Gregg County at night sizing up what had to be done to clean up the county. Now, with the help of a posse he would raid his targets, seldom failing to round up 100 to 200 each evening.

Lacking a big enough jail, Lone Wolf improvised in an abandoned church by installing a heavy chain. Each man was linked to the chain by a smaller chain padlocked around his neck. The "trotline" also served as a method of parading the prisoners along the street for all to see. After checking their records and shipping off those who were wanted or had a criminal record, he gave "suspicious characters" four hours to leave town. Chained to the trotline, fed a watery stew once a day from a can and using a communal bucket for a toilet was a hell of an incentive to leave East Texas and not return.

Respectable oilmen and family men caught in Lone Wolf's raids and paraded through town along with the oil patch trash were released with a warning to stay out of dens of iniquity; however, they probably had a lot of explaining to do to their wives.

SUSPICIOUS CHARACTERS

Lone Wolf Gonzaullas spotted suspicious characters by shaking hands as he made his rounds in the gambling halls and saloons. If a man had soft uncalloused hands and didn't have a job, the suspicious character didn't belong in the oil patch and was put on the trotline.

17
HARD TIMES

Roosevelt was elected President in November 1932. Americans feared they would have little to be thankful for that Depression Thanksgiving. FDR's inaugural address containing the famous line would not be heard for months: "The only thing we have to fear is fear itself."[31]

In East Texas one of their worst fears was realized. The Black Giant was acting like no other oil field. The field didn't contain sufficient gas to drive the oil to the surface and the pressure at the bottom of the wells was dropping rapidly. If it continued to drop, they feared it wouldn't be able to produce oil much longer. At first, geologists, who generally have an opinion about everything, admitted they didn't know the reason.[32] Geology was still not highly respected by oilmen, big and small, who had drilled wildcat wells on a geologist's assurance there was oil in an anticline only to drill a duster. Moreover, most geologists worked for the government or Big Oil; thus, their opinions were tainted as far as the independents were concerned.

Finally, petroleum engineers determined that the oil field was not driven by gas, but by salt water from the west and the underground aquifers feeding the Black Giant were not flowing sufficiently to replace the oil being drained with reckless abandon. Salt water, the remains of an ancient ocean trapped below the surface is heavier than

---

31 A real historian, Richard Hofstadler, swears FDR stole the line from Thoreau: "Nothing is so much to be feared as fear."

32 *See footnote 2.*

oil and found below the oil. Knowledgeable oilmen were aware that the water pressure had to be maintained and allowed to replace the oil as it was produced or most of the oil might not be recovered. While the oilmen, engineers and geologists were discussing what could be done to save Dad Joiner's ocean of oil, the befuddled TRC shut down the entire oil field.

No one was satisfied with the TRC's drastic move. Hot oil increased, abetted by the TRC's underpaid guardians and National Guardsmen slipped a few dollars to look the other way. Families from a one-mule cotton patch, who had not made a cash crop of $500 a year in the decade prior to the boom had tasted the fruits of their oil leases and purchased cars and farm equipment with bank loans they could not repay without their monthly royalty payments. Their roller coaster ride had taken them up to heaven then dropped them into the abyss of hell made worse by the exorbitant boom prices, strangers taking over their quiet towns and the ominous shadow of Big Oil. Warning shots were fired and oil pipelines dynamited.

When the TRC finally opened the field, it set the well allowable at 28 barrels a day, far below the break-even point for the small independents, although the price had crept up to 85¢ a barrel. But even that was short-lived. The court invalidated the TRC's prorationing order because it was based on a per-well criteria, and the wells were again shut-in.

Of the 19 prorationing orders issued by the TRC for the East Texas field during 1932, every one was eventually ruled invalid by the courts.

With H.L.'s wells shut-in or the allowables so low they barely covered expenses, Dad's oil payments were far below what he had anticipated. Dad had spent the $800,000 H.L. paid him over the last two years on wildcat ventures and high living. As usual he was broke when he met with H.L. to hit him up for another advance. Although bedridden with a back injury incurred when he and another man lifted a car off a victim of an accident, H.L. was anxious to see Dad because of nasty rumors floating around that Dad was going to sue him for fraud in the purchase Dad's leases.

After chatting for two hours about everything but Dad filing a law suit, H.L. asked him about the rumor. Dad responded by putting his arm around H.L.'s shoulder, his eyes moistening with tears, and whispered, "Boy, I'd never do anything like that, I love you too much."

H.L. called it intuition. Others said, "It takes one to know one." H.L. related that he had a hunch when the old man put his arm around him and there was a tear in his eye. With only two days before the statute of limitations would run on the case, he knew the sentimental con man was going to sue.

H.L and his attorney worked through the night and the next day with one of his major equipment suppliers drafting a $3 million mortgage and lien on the leases obtained from Dad. Although the supplier did not need or desire a mortgage, he wanted Hunt Oil's continued business. If Dad sued, the liens would take precedence and prevent Dad from attaching the leases. H.L.'s lawyers recorded the mortgage one hour before Dad's lawyers filed the case of *C.M. Joiner v. H.L Hunt, et al.* for fraud, demanding $15 million or the return of the leases. To show there were no hard feelings, Dad sent H.L. a crate of pink grapefruit at Christmas.

Twas the season for law suits, when lawyers celebrate Thanksgiving. Dad sued Ed Laster for fraud in connection with Ed's deal with Mid-Kansas Oil & Gas Co. and disclosure the Daisy Bradford No. 3 had hit the Woodbine sand. Ed counter sued Dad for 360 acres of leases he had earned as a driller. Dad and Ed settled out of court.

Ed threatened to sue Mid-Kansas because it had not assigned him the 1/4 interest in leases of 1,100 acres it acquired for $1 an acre on Ed's tip, now worth $3.3 million. Ed settled for a small fee and another lease.

The most interesting suit from a romantic's view was *Stella Sands v. Columbus Joiner*, another case in which Judge R.T. Brown proved worthy of the title, the Sage of East Texas. Spinster Stella produced love letters in which Dad had promised her half of his ocean of oil. The evidence was not enough to constitute a partnership, but Judge Brown thought the few thousand dollars she sent Dad were worth $20,000 and ordered him to pay the sobbing woman.

Happy to see 1932 come to a close were the men of the Texas National Guard, now able to go home to spend Christmas with their families after the Supreme Court ruled Governor Sterling's martial law was a sham and unconstitutional. Texans have always been a proud bunch and never afraid to do their duty. They whipped Santa Anna and the Mexican Army, fought valiantly for the Confederacy and contributed more than their fair share of Rough Riders in Cuba and to the American Expeditionary Forces in World War I. However, it stuck in their craws to have to go against a fellow Texan...a brother fighting to earn a living.

The troops did not realize that the most famous of all Texas born soldiers and no doubt the most beloved was now in the Army serving his country — Dwight David Eisenhower from Denison, Texas. Like his fellow Texans serving with the National Guard in East Texas, Ike had a dirty and thankless job to face in 1932. As a major, he served as an aide to General Douglas MacArthur, who led the United States Army against hungry and homeless World War I veterans.

Veterans, calling themselves the Bonus Expeditionary Force, had marched on Washington begging for the early payment of their veterans bonus to feed their families during the Depression. President Hoover refused to meet with their representatives, believing they were trouble-making "Reds." MacArthur defeated the unarmed 25,000 veterans with tear gas, bayonets and cavalry commanded by Major George M. Patton, routing them in the infamous "Battle of Pennsylvania Avenue." That night, MacArthur ordered the leveling and burning of their shantytowns of tents, cardboard and packing crates by tanks and cavalry, scattering screaming women and children with tear gas, and injuring hundreds. Two babies suffocated and died from the tear gas. America was at its lowest ebb.

According to no less an authority than my father, there was damn little to be thankful for in 1932. He also told me that Herbert Hoover was a son of a bitch.

TEXANS IN THE WHITE HOUSE

Hoover lost the presidency to Roosevelt, who had the good sense to have a Texan, John Nance Garner, as Vice President, although Garner didn't think much of the job. He said being Vice President, "Ain't worth a pitcher of warm spit."

~~Two other Texans,~~ Lyndon Johnson ~~and George Bush,~~ didn't pay attention to Garner and used the job to become President. [George Bush was born and raised in Connecticut, which any Texan can tell from his Yalie voice.]

PART FOUR

HISTORY

> *Most of the great results of history are brought about by discreditable means.*
>
> Ralph Waldo Emerson

18
DOLLAR OIL

Roosevelt's most colorful Cabinet appointment was Secretary of the Interior Harold L. Ickes. It is not unfair to describe Ickes as a cantankerous curmudgeon. The power-hungry Republican lawyer coveted the reputation during in a Democrat administration in his memoirs, *Autobiography of a Curmudgeon*. The tyrant ran the Department of the Interior with wiretaps and spies. Periodically, he locked the Department's doors at 8:00 a.m. and waited by the door to chastise and fine late for work bureaucrats.

Unheard of in politics, Ickes endured as Interior Secretary through FDR's thirteen years in office and a year into the Truman Administration when Harry realized that one curmudgeon in Washington was more than enough. During his tenure, Ickes convinced Roosevelt and Churchill to sign the Anglo-American Petroleum Agreement of 1944, which failed ratification when the Senate realized the leaders of the free world were planning to split up the vast oil fields of the Mid East, as if the Arab nations were colonies. He also put the bug in Truman's ear to declare that all waters within the three-mile limit of the continental shelf, believed the territory of the states, belonged to the federal government and placed it under the management and control of the Interior Department. After Harry vetoed two bills passed by Congress returning the oil-rich waters to the states, Eisenhower signed the third bill passed by Congress, making Texas, Louisiana and California oil-royalty wealthy again.

Ickes realized America's oil dependence: "We have passed from the stone age, to bronze, to iron, to the industrial age, and now to an age of oil. Without oil, American civilization as we know it could not exist."

He was concerned that the oil industry was wasting oil and teetering on the edge of collapse. Crude oil at 10¢ a barrel was far below its cost to produce, particularly in the Mid West and small oil fields where production costs ran 80¢ a barrel.

The federal government also had a selfish interest. It owned one-third of the nation's lands, including one-half of the oil lands in the Western states. The Interior Department, charged with the management of the public lands, was the largest collector of oil and gas royalties in the country and the Roosevelt administration badly needed money. Like the farmers in Texas, Ickes had no difficulty figuring that a royalty of 1/8 of $1 was ten times more than 1/8 of 10¢.

As Secretary of the Interior, Ickes became America's oil czar with his second appointment as Oil Administrator under the National Recovery Administration (NRA).[33] The newly created NRA was part of Roosevelt's National Industrial Recovery Act (NIRA); however, Ickes paid scant heed to the NRA's Administrator, General Hugh Johnson, who Ickes said was afflicted with "mental saddle sores" and "halitosis of intellect."

As with any new administration, advice rolled in like a tidal wave. Walter Teagle of Standard Oil of New Jersey and independent oilman Harry Sinclair urged Ickes to set federal price and production controls. Aware Sinclair had been involved in the bribery of Interior Secretary Albert Fall in the Teapot Dome scandal, Ickes told friends that he trusted him as much as he did a "snake in the grass," which he equated with everyone with Big Oil.

---

[33] Busybody Ickes third appointment was Public Works Administrator, charged with constructing buildings, roads and projects on government lands to generate employment. One of his first projects was a new building for the Department of the Interior, which still stands on the two blocks between 17th and 19th Streets, N.W. and C and D streets in Washington. It continues to provide the Secretary of the Interior with one of the most palatial offices of any Cabinet member, complete with a private elevator from the basement so the sneaky old curmudgeon's lady friends could visit unnoticed.

Teagle, however, had an intriguing idea. He suggested the East Texas oil field be shut in and purchased as a Naval Petroleum Reserve. As Humble Oil owned 13% of the Black Giant, it could provide a handsome profit and get rid of the pesky independent oilmen. Fortunately, the federal government was broke and Ickes couldn't figure a way to place the East Texas field under the Interior Department rather than as a Navy oil reserve, or the United States would have had the Republic of Texas as its southern neighbor instead of Mexico.

In May 1933 oil prices in East Texas reportedly plunged to 4¢ a barrel. In July Roosevelt issued an Executive Order authorizing the NRA to stop the shipment of hot oil based on a plea from the TRC chairman and Interior Department estimates that hot oil was running at 500,000 barrels a day.☺ Ickes ordered fifty federal agents to East Texas and sent telegrams to the governors of the oil-producing states advising them of each state's production quota.

The monthly forecasts of oil demand and quotas were made by the Bureau of Mines, which had been going through the futile exercise since 1930 at the urging of the API. Actually, the Bureau of Mines number crunchers had no realistic idea how much crude oil was being produced or consumed. In most cases it was being spoon-fed domestic oil crude oil production, storage and consumption figures by the Big Oil controlled API. Even Washington bureaucrats were aware they had no way of telling how much crude oil the East Texas hot oilers and their Oklahoma cousins were producing, nor were the refiners processing the hot oil, large and small, about to report their illegal activities.[34]

Federal agents combed the oil fields and inspected storage tanks, refineries and pipelines, but hot oil continued to flow, much like bootleg booze flowed during the G-men's Prohibition crackdown. The Fed's principal enforcement tool was a regulation by the Federal Tender Board

---

34 The Bureau of Mine's ineptitude in forecasting the monthly demands has been laid to the inexperience and political bias of the administration. It consistently underestimated demand until late 1938, then overestimated until 1941. During World War II every oil field produced its maximum to fuel the war effort. After 1945 the Bureau's underestimate of demand by as little as 2% caused spot shortages around the nation. This worked well for the majors who owned most of the storage facilities and could help alleviate a shortfall and make an extra buck during price bumps caused by the scarcity.

requiring all oil be accompanied by certificates proving the black goo was produced under the NIRA allowable. As mentioned earlier, crude oil is fugacious, it migrates underground. It is also fungible and can be freely exchanged. On the surface there was no way the G-men could tell the difference between federally fungible certified oil and hot oil as it migrated across the country in trucks, trains and pipelines.

The stream of hot oil was aided by corruption and incompetence. The majority of the federal agents were political hacks with little or no experience in the oil field. The going rate for a Federal Tender Board certificate from a federal agent or off a Dallas printing press was $5, including the necessary signatures of federal officials, sometimes bearing the signatures of such notables as *Jeff Davis, R.E. Lee, Hal Ickes, Frank Roosevelt and Ed Hoover.* Texans found it was easy to bribe a federal agent, something they couldn't do with a Texas Ranger.

Nevertheless, Ickes' statistics showed that hot oil was drying up and, as if by magic, the price of crude oil crept up to $1 in 1934. Part of the credit, however, should be given to Congress' levying an import tariff on crude oil of 21¢ a barrel and $1.05 on refined petroleum products.

The tariff hit Venezuela hardest as it was exporting 55% of its oil to the United States. However, the international oil companies shipped their cheap Venezuelan oil to Europe and it soon surpassed the United States as Europe's largest supplier. Mexico wasn't as fortunate. Much of its crude oil was inferior heavy gunk and in little demand by refiners who required lighter crudes to manufacture gasoline. The left-wing Mexican government, unhappy with the drop in exports, Big Oil's shenanigans, and its inability to pay the wage increases promised the bloated work force of the politically powerful oil field workers, nationalized the oil industry and tossed Big Oil out of Mexico. Big Oil had more oil than it could sell, but was chagrined that the ignorant peasant republic wouldn't pay for the equipment it had expropriated. When Big Oil complained to Roosevelt, he laughed and said he couldn't help because of his "Good-Neighbor Policy."[35] Mexico's national oil com-

---

[35] FDR stopped laughing when Mexico became Nazi Germany's largest oil supplier and the Japanese were exploring for oil south of the border. During World War II, the United States blockaded German tankers and bought a few barrels from Mexico to keep the peasants happy, after all, we were good neighbors.

pany, Petróleos Mexicanos (Pemex) learned something from Big Oil. In no time, Pemex became the world's most corrupt national oil company. (It still is.)

Big Oil had a few more tricks up its sleeve to keep prices and profits up. In fact, the national average price of gasoline at the pump increased from 17.0¢ to 18.9¢ per gallon between 1931 and 1935 according to API statistics.[36] Eastex gasoline refiners, such as Hunt Oil's Parade, numbered 75 in 1934[37] and independent gasoline stations could sell cheap Eastex for as low as 10¢. Competition to sell gasoline was described as a "jungle of throat-cutting." The majors and large independents enticed motorists by giving away trading stamps, ashtrays, cigarette lighters and glasses. The Eastex boys, not to be outdone, handed out practical items, such as a dozen eggs and chicken dinners with a fill up.

The competition at the pump engendered Big Oil's worst un-American conspiracy, which they had the audacity to set in writing. Their *Draft Memorandum of Principles* dated January 1, 1934, restricted advertising and claims of gasoline superiority. Today, that would mean Exxon couldn't "put a tiger in your tank." Newspaper ads and billboards were restricted and the size of signs at dealer stations were limited. American "give-away" institutions — free glasses,[38] dishes, cigarette lighters, key chains and road maps were curtailed. The sponsorship of racing cars and the plastering of cars and drivers with logos was cut, which would have left the Richard Pettys and Al Unsers of the times naked without their STP and Valvoline caps and patches.[39]

---

36 The average gasoline price dropped from 20.0¢ between 1930 and 1931, attributed to price wars and one-half million less cars on the road at the beginning of the Depression, but before the East Texas field began producing.

37 No one is sure of the exact number of East Texas refineries. When a federal agent suspected a refinery outside of Kilgore was refining hot oil, he discovered a good ol' East Texas boy had built a hellacious big still and was cranking out barrels of bootleg whiskey.

38 If my mother knew about their jiggery-pokery, she would not have forced my father to stop at the Shell station to get a free dish or glass on the way to town. I still treasure my dad's Shell glass he sipped an occasional lemonade from.☺ (I lied. It was Scotch.)

39 Mention of the Indy 500 winners, Bill Cummings (1934) and Kelly Petrillo (1935) would not have been as meaningful except to a few old timers. Most of today's drivers have switched to better deals offered by Marlboro, Budweiser, Domino's Pizza and other great American enterprises.

Ickes, unaware the Big Oil scoundrels were about to deprive Americans of their "freebies," welcomed Big Oil officials to help draft provisions of the NIRA Oil Code geared to eliminate price wars at the pump they claimed to be crippling the oil industry. The foxes invited into the chicken coop added prohibitions against paying dealers for advertising on the dealers' premises and repairing and painting the stations (except pumps), which would not only save Big Oil money, but put the task of renovation and painting on the back of the little dealer.

Price maximums were also set under the NIRA Oil Code. Dealers were required to post prices and forbidden to change the price for 24 hours or sell below the posted price in an effort to stop price cutting. (That's when the cute little "9/10" of a cent you can barely see after the large price numbers at your service station originated.) As an example, guidelines were established in areas making it illegal to sell a barrel of crude oil at less than 18 1/2 times the gallon tank car price (wholesale) of 60 octane gasoline, which was below the lowest grade of gasoline sold by the majors and generally what the Eastex refineries were cranking out.[40]

But that didn't stop the Eastex boys and Okies from selling below the majors and large independents. When the competition became unbearable, the majors decided on a sure method to cut off the problem. The majors and large independents in East Texas and the Mid-Continent[41] agreed to take "dancing partners." Translated: they stemmed the flow of cheap gasoline by purchasing it from a small refiner dancing partner who had been selling Eastex gas to independent distributors and stations. This allowed the small refiner a greater profit and raised the price of gasoline at the same time. Some majors "married" their dancing partners by buying them out and obtaining an agreement from the small refiner never to go dancing again.

---

40 The formula is too complicated for this irreverent tale. Every refinery is structured differently and is saddled with varying operating costs. It is enough to say that the regulations worked out to crude oil selling roughly at $1 a barrel. The average octane ratings were 70 for regular and 76 for premium in 1935, far below the 94 octane super premium sold today.

41 The oil industry generally refers to the Mid-Continent as everything between the Mississippi River and the Rocky Mountains, but excluding the troublemaker, East Texas. Most dancing partners were located in Texas, Oklahoma and Kansas.

Of course, Big Oil members cheated on each other. The sneaky advertising and marketing provisions of the Oil Code were almost impossible to enforce. It took World War II to stop Americans from demanding trading stamps or free dishes at their local stations. Thus, the marketing provisions of the NIRA Oil Code never amounted to pimple on a mule's butt, as they say in East Texas, but showed what sneaks the whole bunch were.

In 1939 the Supreme Court ruled that a combination of major oil companies violated the Sherman Antitrust Act by conspiring to fix oil prices in the East Texas and Mid-Continent oil fields.[42] The defense of the 16 companies and 30 individuals finally tried in court, after years of legal finagling ended by dismissing 11 other companies and 26 individuals, turned out to be that they were working under the NIRA Oil Code when they took on dancing partners and purchased the Eastex gas from the small refiners *and* that the Oil Administrator, Harold Ickes, knew about their scheme. However, tricky Ickes never gave Big Oil permission in writing and refused to testify, proving he was a smart curmudgeon. He may also may have been a loveable curmudgeon. In the middle of oil crises while in his seventies he married a woman forty years younger.

---

42 *United States v. Socony Vacuum Co.*, 310 U.S. 150 (1939). Law students studying antitrust law cut their teeth on this dry (117 page) case holding that any combination to fix prices is a *per se* violation of the Sherman Antitrust Act. But how many students are aware that Chief Justice Hughes had to recuse himself from the case because the former API lawyer knew all the culprits?

CHEATERS

Sam Rayburn, former Speaker of the House of Representatives from Texas was quoted: "Any fellow who will cheat for you, will cheat against you." Sam was right. The majors sold gasoline at 9 1/2¢ a gallon and handed comic books and children's games to my brother and me sitting in the back seat of Daddy's Ford to keep us quiet.

By the time 1935 rolled around, the flow of hot oil had slowed to around 35,000 barrels a day according to the Washington paper shufflers.© There were less that three dozen teakettle refineries left in the East texas field. The majors and large independents owned 60% of the wells, which meant the small independents had been able to hold on to 40%. Humble's 1,500 wells made it the largest producer. Hunt Oil owned 250 wells and was the largest independent in the East Texas field ...not bad for a poor boy. The best news was prorationing tied to the meaningful conservation of the Black Giant's oil began to work relatively fairly and the price of crude oil had hovered around $1 a barrel for a year. But that didn't mean that the battle was over.

Attorneys for the Department of Justice discovered they were holding a bear by the tail when they appeared before the Supreme Court in December 1934 and faced a country lawyer in a formal cutaway coat and stripped pants. The case was of such importance the court scheduled two days for the oral argument. East Texans claimed his argument before some of the greatest minds to sit on the high court, Brandeis, Stone and Cordozo, "turned the Department of Justice and Ickes every way but loose."☺

Fletcher "Big Fish" Fischer had "read the law" in the Oklahoma Territory as a young man while toiling on his daddy's farm. He arrived in Tyler, Texas, during the early days of the oil boom and opened his law practice by representing independent producers and refiners from an office above a gasoline station. With little formal education, his language was of the country, ungrammatical and chock full of metaphors. The name "Big Fish" suited the burly farm boy, he was as homely as a catfish, but devoured the Department of Justice defenders of the NIRA Oil Code like a shark.

In an opinion written by Chief Justice Charles Evans Hughes, the API's former lawyer, the Supreme Court ruled the provisions under the NIRA Oil Code unconstitutional. Five months later, the Supremes declared the NIRA in its entirety unconstitutional.[43]

Many independents opened up their valves and let the oil flow full blast at the news that Big Fish had kicked butt in Washington. Again, the victory was short-lived. Senator Tom Connally of Texas, a strong backer of "dollar oil," asked Interior Department attorney, J. Howard Marshall, to draft a bill that was to become known as the Connally Hot Oil Act. The new law prohibited the interstate transportation of oil produced in violation of state conservation laws and carried stiff penalties and jail terms.

Congress followed up by the passage of the Interstate Oil Compact Act of 1935, permitting the oil producing states to coordinate and adjust the Bureau of Mines demand forecasts, not regarded as worth a damn by the states. As expected, a government that couldn't prevent the sale of booze, couldn't stop the flow of hot oil entirely. After the discovery of large pools of oil in Illinois, the state refused to join the Interstate Oil Compact until it reaped its ill-gotten largess, which created snags for several years and irritated Texas and Oklahoma who had to reduce their production.

---

43 Big Fish's victory may be found in *Panama Refining Co. v. Ryan*, 293 U.S. 388 (1935). Every law student has spent hours in law school discussing the *Schechter Poultry* "sick chickens" case declaring the NIRA an unconstitutional delegation of power to the President to regulate interstate commerce, so no citation is necessary.

WHO WAS J. HOWARD MARSHALL?

Marshall was an Interior Department lawyer in charge of the enforcing the NIRA Oil Code in East Texas and credited with formulating many oil regulations, creation of Federal Tender Board certificates and drafting the Connally Hot Oil Act. Armed with knowledge of loopholes and oil industry wheeling and dealing, he moved to Texas and amassed an oil fortune of $725 million. Few outside Houston heard of Marshall until, at the age of 90, he married Texas model Anna Nicole Smith, one-third is age, then promptly kicked the bucket.

To show its independence, the TRC periodically exceeded the Interstate Compact quota by a "smidgeon" and amended the Marginal Well Act to exempt wells producing less than ten barrels a day, which every state eventually followed. That didn't mean the East Texas oilmen, big and small, were satisfied with the TRC's allowables. On many occasions the TRC's allowables were challenged successfully in court.

Today, oilmen in the East Texas oil field recognize that the Black Giant must be nurtured and follow proration rules limiting production to a maximum efficient rate. However, the issue was not settled until "dollar oil" was reached and maintained through witchcraft and the demand for oil during World War II temporarily suspended the evil of economic waste and market demand prorationing.

Before the hot oil era of the 1930s came to an end, over 100,000,000 barrels of hot oil had gushed out of the Black Giant (4,200,000,000 gallons, if you are an environmentalist). But that's not to say that prorationing monkey business didn't continue to take place in the oil patch by both the big and little guys. In 1950 Humble oil was convicted of violating the Connally Hot Oil Act.[44]

---

44 *Humble Oil & Refining v. United States,* 198 F.2d 753 (10th Cir. 1952.) In 1985 Humble Oil's pricing practices caused its parent corporation, Exxon, to be found guilty of overcharging on crude oil prices from the Hawkins oil field in East Texas under Department of Energy price control regulations. Exxon was ordered to make restitution of $1.6 billion plus interest. *United States v. Exxon Corp.,* 773 F.2d 1240 (TECA 1985).

19
THE BOTTOM OF THE BARREL

In refining crude oil, after gasoline, jet fuel, diesel fuel and other valuable petroleum products are distilled, what remains is heavy gunk, "residual oil," or in oilmen's jargon — *the bottom of the barrel.* The ending of the story is analogous. Many took out what they believed valuable during the East Texas oil boom and left the bottom of the barrel.

The tales of Dad Joiner, Doc Lloyd and H.L Hunt are an amalgam of fact and legend. The sordid behavior of Humble and Big Oil has been documented in the congressional hearing and reports listed in the Bibliography. Unlike most boring, self-serving tomes Congress disgorges, parts read like intricate plots from a Tom Clancy or John Grisham novel. To the reader, two questions should become apparent: What took Congress and the White House so long to figure out what the hell was going on? And did our Presidents and Congress really know all along?

But there were also good guys. There were many East Texans like Malcolm Crim, Walter and Leota Tucker and banker R.A. Motley — hard-working and honest. There were also many small independent oilmen who survived — all hard-working, but maybe not so honest. It is not for me to judge. I would love to believe all the tales are true, but we all know Texans tend to exaggerate. What is true is that out of Texas evolved that odd breed — a combination of populist and conservative — who had tasted both poverty and wealth, but never forgot their roots in the soil and what is was to struggle for what they got. Rough around the edges, many skirted the law — unfair and unconstitutional laws — in their battle "agin' Big Oil* and big gov'mint" [*pronounced "awl"], which they couldn't trust and still lies at the root of their mistrust today.

Unlike the heavy muck at the bottom of the barrel, they were distilled from the top of the barrel — *the light ends* — the gasoline and jet fuel that drives the nation.

Big Oil grabbed the king's share of the Black Giant, but a passel of the small independents survived and became "Texas rich." Big Oil became bigger. Look at your choices when you fill up your car: the former Standard Oil Companies — Exxon, Amoco, Mobil, Conoco, Ashland, Marathon, Chevron (it swallowed Gulf) and Arco (it snatched Sinclair) — and the other Seven Sisters — Shell, BP (it devoured Standard Oil of Ohio) and Texaco (it grabbed Getty.) As this tale goes to press, BP is taking over Amoco and Exxon and Mobil are merging.

A 1947 Federal Trade Commission report bearing the sinister title *The International Petroleum Cartel* revealed the corrupt machinations of the Seven Sisters, but was classified and hidden in the bottom of the barrel by the Truman administration. "Sanitized" and finally released under political pressure in 1952, it became a best seller in places like Saudi Arabia, Iran, Iraq, Kuwait and Venezuela. The countries read what they already suspected: they were being screwed by Big Oil. In 1960 the five major oil exporters formed their own oil cartel, the Organization of Petroleum Exporting Countries (OPEC).

OPEC

Contrary to myth, OPEC is not a bunch of greedy Arabs. Iranians, Indonesians, Nigerians and Venezuelans are not Arabs. The small nations formed OPEC after learning how big oil operated. Part of their education included visiting the Texas Railroad Commission to study market demand prorationing.

OPEC was the concept of Juan Pablo Pérez Alfonzo, Venezuela's left-wing Minister of Mines, who resigned after discovering that oil was corrupting OPEC, including his native Venezuela. In his memoirs, he labeled oil "the excrement of the devil."

The co-founder of OPEC, Abdullah Tariki, was an anti-Western, leftist nicknamed the "Red Sheikh" and Saudi Arabia's first modern oil minister. Tariki began

> his dislike of Americans while studying geology at the University of Texas and working for Texaco. He claimed Texans tossed him out of bars because he looked Mexican. Tariki was canned from his job and exiled from Saudi Arabia after he objected to a Saudi prince receiving a kickback on an oil concession, proving he didn't really know how Big Oil or his native land did business...with *baksheesh.*

In 1952 Truman finally ordered the Justice Department to criminally prosecute the Seven Sisters (Exxon, Mobil, Chevron, Texaco, Gulf, BP and Shell and the French stepsister CFP) under the United States antitrust laws, then had second thoughts eight days before he left office and told the Attorney General to reduce the case to a civil action.

Under Eisenhower the charges against the foreign sisters, BP, Shell and CFP, were dropped[45] and the Americans charged with "an unlawful combination and conspiracy to restrain interstate and foreign commerce." The case never went to trial. The American companies entered into a consent decree in which they neither admitted or denied the charges and promised never to do it again.☺

Why did Harry and Ike fail to go after Big Oil? It was the advice of both their Secretaries of State and Defense, Directors of the Central Intelligence Agency and the National Security Council that prosecution would be contrary to the "national interest." Big Oil got away with what it did in East Texas and other parts of the world, including the gas pump where you fill up your car, in the interest of "national security." Big Oil had been instrumental in assisting the government in everything from overthrowing heads of state, squeezing little obnoxious oil

---

[45] The foreign companies were not prosecuted, in part, because the British, Dutch and French governments owned big chunks of their stock.

nations dry and double-crossing so-called friendly countries, and the State Department and CIA didn't want to let the cat out of the bag. Also, Big Oil was handy when election time came around.[46]

OIL & POLITICS

"The trouble with this country is that you can't win an election without the oil sector, and you can't govern with it."

Franklin Delano Roosevelt

Big Oil's legacy in Texas, America's oil capital, is distrust. Texas remains the only oil producing state not to have legislated compulsory unitization. Henry L. Doherty's "rationalization" that oil fields should be nurtured cooperatively — unitized — so we are able to reap the maximum amount of the precious resource, cannot be required under Texas law because of Texans' wariness of Big Oil and big government. Today, roughly half the oil produced in the United States comes from fields partially or wholly unitized to permit the secondary recovery of oil by means of water flooding or gas injection. In Texas, operators may voluntarily unitize their fields, but Big Oil and government cannot make them, thanks to Big Oil's legacy.

The *Joiner v Hunt* lawsuit was not the barroom brawl people expected. Rumors flew around Henderson, Houston and Dallas of fraud, bribery and intimidation. Frank Foster, who drilled the Deep Rock well, and Charlie Hardin, H.L.'s oil scout who Foster kept informed of the well's success in hitting the Woodbine, could not be found to be served with a subpoena. H.L.'s memory failed him during the depositions when asked why he gave Foster $20,000.

---

46 For the anti-Big Oil view reading like an indictment, read John M. Blair's *The Control of Oil*. Blair was the principal drafter of the Federal Trade Commission's *The Staff Report of the International Petroleum Cartel*. A more balanced picture is given in Daniel Yergin's Pulitzer Prize winner, *The Prize, The Epic Quest for Oil, Money & Power*.

H.L. admitted he knew of the Deep Rock well's success three or four hours before he signed the agreement and insisted he had told Dad. Dad swore that he had been bamboozled and kept in the dark about the Deep Rock's well. Neither wanted to stake all or nothing on the wisdom of twelve men and women weighing the testimony of a gambler on one hand and a promoter who had sold his interest in the field three times over on the other hand. There was also the rulings and irascible wisdom of Judge Brown, the Sage of East Texas, to be considered.

No one knows what was said or agreed between the two wheeler-dealers after they went behind closed doors to settle the lawsuit. Dad withdrew his suit before a packed courtroom of newspaper reporters, investors and curious East Texans left to speculate what happened after the two legends walked out of the courtroom.

The author will not speculate what Dad received, if anything. The two oilmen remained friends and Dad never complained. Dad continued to receive advances on the oil payments from H.L. along with H.L.'s admonition that a true wildcatter never gambles his own money on a wildcat well. It was reported that a month after the trial Dad asked H.L. for a $10,000 advance to pay his attorney's fees in the suit, but H.L. insisted he advance Dad only $3,000 because, the lawyer lost the case.

Pregnant Frania Hunt got wind that there was another Mrs. Hunt in 1934 and moved to Great Neck, New York, where she bore H.L. another son, Hugh (or Hue?). Two years later he sweet-talked Frania into moving to Houston, where she again became the wife of Major Franklin Hunt. Becoming a well-known, wealthy oilman had it drawbacks. It was difficult for H.L. to have a wife in both Dallas and Houston, so he attempted to convince her to move to Utah and become a Mormon. "Out there, having two wives is normal in that religion," he pleaded to no avail. Although H.L. was a suave con man, he was not a poet, as evidenced by the following doggerel he sent Frania in 1937:

Fran, when flying over the mountains
To the land of Brigham Young
She had done a little flitting around
And by the love bug she was stung.
She joined a manly man
A Mormon to become.

Frania Tye (Tiburski) Hunt, a nice Polish Catholic girl from Buffalo, New York, had no choice but to walk away from H.L.'s country-boy charm. In 1942, H.L., now a multi-millionaire, forced a cash settlement on Fran of $100,000 plus $2,000 a month income from one of his East Texas oil leases in turn for her relinquishing all claims against him and stating that they were never legally married. Later, Frania and the four children, Howard, Haroldina, Helen and Hue, would contest the settlement, but that's another story a myriad of writers have already squeezed dry of blood.

H.L. Hunt became a billionaire and right-wing curmudgeon. There are no typical billionaires, there are too few, but he did develop some traits of the very rich in his later years. No longer a "spiffy" dresser, he became "tight as a tick" with money and wore out shiny suits bought off the rack, but was known to bet $100,000 on a college football game. He gave up booze and became a health food fanatic, disdaining white flour. My one invitation to lunch with him in 1964 was an example of his daily fare. I dined in his Spartan office on a cheese sandwich (one slice of cheese) on wheat grain bread and carrot sticks and discussed oil and politics. However, at night he went to his Dallas mansion, a replica of George Washington's Mount Vernon, only bigger.

H.L. met his demise at eighty-five, leaving his fourteen living legitimate and not so legitimate children and two remaining wives to squabble over a couple of billion dollars, attempt to corner the silver market and get into all sorts of trouble.

Throughout his life H.L. was haunted by rumors and accusations that he swindled Dad Joiner. H.L. called it his "greatest business coup." H.L.'s friends claimed he merely conned a con man.

EVERYONE'S FAVORITE H.L. HUNT TALE

Lamar Hunt is the best known and respected of the H.L.'s offspring. He was graduated from Southern Methodist University with a degree in geology, which may explain why the Hunt Oil Co. finally established a geology department.

As a third string end on SMU's football team, Lamar acquired the nickname "Poor Boy" and a love of football. The name had nothing to do with the way H.L. and Dad drilled oil wells in the 1920's. His father had become one of America's richest men. Denied playing the sport he loved, like any rich man, he became a founder of the American Football League and bought the Dallas Texans, which became the Kansas City Chiefs.

When a newspaperman asked H.L. what he thought about Lamar losing $2 million the first year of the football franchise and $3 million the second year, H.L. replied: "That's terrible, at that rate he'll be broke in 250 years."

Dad and Dea England ran off to Juarez, Mexico, in 1933 where Dad got a quickie Mexican divorce from his wife of over fifty years and Married Dea. Dad was seventy-three and Dea was twenty-five.

Dad wandered throughout Texas looking for another ocean of oil until he was eighty-seven, but only drilled dry holes. He passed away in Dallas in 1947...broke. Some folks said it was because he let Dea have a charge account at Neiman Marcus.☺ However, he will always be remembered as *The Daddy of the East Texas Oil Field.*

Dad's ocean of oil — the Black Giant — still spouts oil. It can no longer gush the 500,000,000 barrels it contributed in four years to the defeat of Nazi Germany and Imperial Japan while connected to the "Big Inch" pipeline running 1,400 miles to Big Oil's Pennsylvania and New Jersey refineries. However, it is still a mighty force in producing jobs in East Texas and reducing America's dependence on foreign oil.

Market demand prorationing brought about after Dad found too much oil is no longer required. It endured over four decades, until the Arab embargo of 1973, when Americans woke up and discovered they had been dependent on foreign oil imports since 1948. America guzzles 25% of the world's oil produced every day, but we have less than six percent of the world's reserves. *Today, the United States imports over one-half of its oil needs — America is running out of oil.*

The Daisy Bradford No. 3 was still pumping crude for the Hunt Oil Co. when I visited the East Texas oil field in 1996 — sixty-six years since Ed and the crew brought in the first East Texas gusher.

As I sat gazing at the well, now on a pump that coughs up a couple of barrels a day, the images of Dad, Doc, Daisy, Walter, Leota, Ed and Dennis hovered about the rickety wooden derrick. I swear I heard Dad Joiner say, "Boy, you ain't seen nothing yet...there's an ocean of oil down there."☺

blank before bibliography

BIBLIOGRAPHY

Anderson, Robert O. *Fundamentals of the Petroleum Industry.* Norman: Univ. of Oklahoma Press, 1984.

Baker, Ron. *A Primer of Oilwell Drilling.* 4th ed. Austin: Univ. of Texas.

Ball, Max W., Douglas Ball, and Daniel Turner. *This Fascinating Oil Business.* New York: Bobbs-Merrill, 1965.

Blair, John M. *The Control of Oil.* New York: Pantheon, 1976.

Boatwright, Mody C. *Folklore of the Oil Industry.* Dallas: Southern Methodist Univ. Press, 1963.

Boatwright, Mody C., and William T. Owen. *Tales From the Derrick Floor.* New York: Doubleday, 1970.

Boller, Paul F., Jr. *Congressional Anecdotes.* New York: Oxford, 1991.

Brown, Stanley H. *H.L. Hunt.* Chicago: Playboy Press, 1976.

Buckley, Tom. "Just Plain H.L. Hunt." *Esquire.* (January 1967), pp.64-69, 140-154.

Burst, Ardis. *The Three Families of H.L. Hunt.* New York: Weidenfeld & Nicolson, 1988.

Clark, James A., and Michel T. Halbouty. *Spindletop.* New York: Random House, 1952.

_____. *The Last Boom.* New York: Random House, 1972.

Davis, David Howard. *Energy Politics.* 2d ed. New York: St Martin's Press, 1978.

Davis, Kenneth C. *Don't Know Much About History.* New York: Avon, 1990.

Fox, William F., Jr. *Federal Regulation of Energy.* Colorado Springs: Shepard's/McGraw-Hill, 1983.

Harter, Harry. *East Texas Oil Parade.* San Antonio: Naylor, 1934.

Hunt, H.L. *H.L. Hunt Early Days.* Dallas: Parade Press, 1973.

_____. *Hunt Heritage.* Dallas: Parade Press, 1973.

Knowles, Ruth Shelton. *The Greatest Gamblers: The Epic of America's Oil Exploration.* 2d ed. Norman: Univ. of Oklahoma Press, 1978.

Kuntz, Eugene O., John S. Lowe, Owen L. Anderson and Ernest E. Smith. *Oil and Gas Law.* 2d ed. St. Paul: West, 1993.

Larsen, Henrietta M., and Kenneth W. Porter. *History of Humble Oil and Refining: A Study in Industrial Growth.* New York: Harper Brothers, 1959. [A Humble Oil PR flack job.]

O'Connor, Richard. *The Oil Barons: Men of Greed and Grandeur.* Boston: Little Brown, 1968.

Petzinger, Thomas, Jr. *Oil & Honor: The Texaco-Pennzoil Wars.* New York: Putnam, 1987.

Rosenbaum, Walter A. *Energy Politics, and Public Policy.* 2d ed. Washington: CQ Press, 1987.

Sampson, Anthony. *The Seven Sisters.* 4th ed. New York: Bantam, 1991.

Solberg, Carl. *Oil Power.* New York: Mason/Charter, 1976.

Tait, Samuel W., Jr. *The Wildcatters: An Informal History of Oil Hunting in America.* Princeton: Princeton Univ. Press, 1946.

Weaver, Jacqueline Lang. *Unitization of Oil and Gas Fields in Texas: A Study of Legislative, Administrative and Judicial Policies.* Washington: Resources for the Future, 1986. [Dry, but packed with facts.]

Werner, M.R, and John Starr. *Teapot Dome.* New York: Viking, 1959.

Williamson, Harold F., Ralph L. Andreano, Arnold R. Daum, and Gilbert C. Klose. *The American Petroleum Industry. Vol. 2. The Age of Energy, 1899-1959.* Evanston: Northwestern Univ. Press, 1963. [Funded by the American Petroleum Institute and shows its bias.]

Yergin, Daniel. *The Prize: The Epic Quest for Oil, Money & Power.* New York: Simon & Schuster, 1991. [A definitive, highly readable work for anyone interested in the petroleum industry.]

82nd Cong., 2nd Sess., Senate Small Business Committee. *The International Petroleum Cartel,* Staff Report of the Federal trade Commission. 1952.

91st Cong., 1st Sess., Senate Subcommittee on Antitrust and Monopoly. *Hearings on Governmental Intervention in the Market Mechanism.* 1969.

93rd Cong., 2nd Sess., Senate Subcommittee on Multinational Corporations, Committee on Foreign Relations. *Hearings on Multinational Petroleum Corporations and Foreign Policy.* 1974

U.S. Department of the Interior, Bureau of Mines. *Minerals Yearbook.* Washington: Government Printing Office. Various.

American Petroleum Institute. *Petroleum Facts and Figures.* Washington: API. 1959.

JAMES M. DAY

James Day practiced law for thirty-five years, specializing in international and domestic oil & gas and mining matters. As Director of the Department of the Interior's Office of Hearings and Appeals during the turmoil of the Arab oil embargo, he was awarded the Department's Outstanding Service award.

He has taught Oil & Gas Law and the Regulation of Energy at the Washington College of Law, the American University, for fifteen years and was named the Outstanding Adjunct Professor in 1988.

At present he resides in Arlington, Virginia, where he teaches, does petroleum consulting and writes.

> **TO ORDER:**
> **What Every American Should Know About The Mid East and Oil,**
> **by James M. Day**
> # call 800-729-4131

An Irreverent, Factual Analysis of Middle East Politics & Oil

This guide to the mysterious history and politics of the Middle East will help Americans decipher the latest diplomatic delusions of their President, Congress and the U.S. State Department, untangle the rationale for Palestinian terrorists bombs in the West Bank and Israeli jets strafing civilians in Lebanon, and explain why your sons and daughters may have to return to the region so you can have enough oil.

The Western colonial powers created the boundaries in the Middle East, without regard to the religion, culture or ethnic background of the people, for selfish reasons after World War I...power and oil. America has failed to wield its influence in the region because of its political leaders' myopic support of Israel and failure to understand and/or refusal to admit to the complex regional issues. As a result, America is now part of the problem rather than offering an equitable solution and is in danger of losing its superpower leverage.

Saddam Hussein scoffs at America's impotent threats, Iran sneers at the Great Satan's sanctions and Saudi Arabia has no faith in our promises. A recent Israeli newspaper reported that "Binyamin Netanyahu owns the U.S. Congress." Will tiny Palestine and oil be America's downfall?

Machiavellian perceptive, irreverently witty and an enjoyable read, this book provides invaluable insights into the volatile Mid East and world oil for the serious student as well as the host or hostess at their next cocktail party... including why gasoline will cost over $3.00 a gallon in ten years.

To Order Other Titles Call 800-729-4131

Printed in the United States
203425BV00002B/181-228/A